The Second Culture

The Second Culture

BRITISH SCIENCE IN CRISIS
THE SCIENTISTS SPEAK OUT

Clive Cavendish Rassam

AURUM PRESS

First published 1993 by Aurum Press Limited,
10 Museum Street, London WC1A 1JS

A catalogue record of this book is available
from the British Library

ISBN 1 85410 233 8

1 3 5 7 9 10 8 6 4 2
1994 1996 1997 1995 1993

Printed in Great Britain by Hartnolls Ltd,
Bodmin

CONTENTS

ACKNOWLEDGEMENTS

Although the original idea for this book was my own it would not have been developed without the help of a good many people. First and foremost my thanks go out to Professor Denis Noble and Dr John Mulvey at Oxford University. It was they who encouraged and supported me at a very early stage. Throughout the writing of the book they were immensely helpful both in suggesting lines of enquiry or scientists I ought to see and in correcting one or two mistaken impressions that I may have had. I am also indebted to Professor Noble's secretary, June Morris, for her helpful comments on parts of the book.

My thanks also go out to all those scientists who gave their valuable time, especially when in some cases, their diaries were allegedly booked three years in advance. Such is the life of an international scientist!

My thanks also go to my editor, Piers Burnett, for his wide-ranging suggestions, which have immeasurably widened the scope of this book. Also to my agent, Andrew Lownie, for some interesting points raised during the writing. Another who has assisted, as with my earlier books, in structuring some of my thoughts has been Sheila Evers.

The conclusions in this book are my own. Many of them will be widely shared by scientists, while some will prove more controversial. What is important is that the issues raised in this book are discussed and addressed.

INTRODUCTION

Up to about the middle of the nineteenth century the word 'science' meant the broad body of knowledge, not just the physical and natural sciences. There was still a common intellectual culture in Britain embracing what we would now call the sciences and the arts. Indeed, many celebrated literary figures such as Wordsworth were as much interested in the new scientific discoveries as they were in the contemplation of nature and the human condition. Wordsworth himself spoke of poetry being 'the impassioned expression which is the countenance of all science'. An educated man was expected to be knowledgeable about both arts and science and it was not thought totally odd if women, too, expressed an interest in science as well as literary subjects.

Today there is very evidently a fissure between the arts and the sciences, with the arts ranking as the pre-eminent culture for most people, and the sciences coming a poor second, if they rank at all as culture in some minds. How far we have travelled over the past 150 years. Today few literary people, or other non-scientists, feel any need to apologise for knowing very little about science; in fact many seem to take a perverse pride in their ignorance, regarding science not as a culture but as an anti-culture.

This rift between the sciences and the arts began to develop in the mid-nineteenth century and, despite the herculean efforts of scientists such as Thomas Huxley, the schism widened as the century drew to a close. In France and Germany, meanwhile, technical and scientific education expanded and became a central part of their cultures. But in Britain the increasingly arts-centred education that most children received from 1870 onwards, coupled with increasing specialisation and professionalisation in the sciences, meant that two nations developed. In one camp were the 'generalists', including the growing army of administrators and managers, lacking all knowledge of science, who ran the country's (and the Empire's) institutions; in the other were the scientifically and technically qualified specialists. The result was two elite groups, neither able adequately to communicate with the other.

By the 1960s, despite the role that science had played in winning the Second World War, the existence of what the scientist, civil servant and novelist C.P. Snow dubbed the Two Cultures had become deeply rooted in British society. The consequences, as Snow knew only too well, were potentially serious. For the decisions which had to be made by Britain's administrative and managerial class were increasingly scientific in nature, yet the decision-makers still had little knowledge or understanding of science and were essentially unqualified to make judgements as between one scientific recommendation and another. Even when British governments began to appoint a Chief Scientific Advisor that person still had to fight to be taken seriously. Anyone reading the memoirs of Lord

Zuckerman, a former Chief Scientific Advisor in the 1960s and 1970s, will see the difficulty of trying to infuse some scientific arguments into policy-making at the top. As Sir Solly Zuckerman, he was undeniably a member of Britain's social elite, on friendly terms with high-ranking members of the ruling establishment; but, as a scientist, he still remained an outsider whose views often seemed to carry little weight.

Today, the two cultures are very much with us. Although we live in a world dependent on technologies the scientists behind those technologies have become invisible. John Durant, a biologist and Professor of the History and Public Understanding of Science at Imperial College, London, argues that while we may know about some of the more newsworthy scientific discoveries, we really have no conception about the process that lies behind those discoveries. He says: 'Most people have precious little idea of what scientific research is, how it is done, who does it, where it is done or even why they do it.' This book is an attempt to make scientists a little less invisible, by asking figures from the major fields of science what it is they do, why they do it and how they rank their achievements, which in most cases are considerable.

The scientists presented in the following pages represent the main branches of current scientific endeavour. Most of them have acquired a considerable reputation in their field – usually for a particular invention or discovery – while some are in the course of doing so. About 30% of these scientists are Fellows of the Royal Society, a much-coveted honour for which there is understandably great competition. Each year, the Society elects a handful of scientists from each branch of science, awarding fellowships mainly on the basis of research achievements. Five of the scientists are Nobel prizewinners. Thus those featured here are, in the main, research scientists, working in universities or research institutes; which is not to say that many of them are not also involved in exploring the applications of their ideas. The others come from industry and management consultancy. Perhaps inevitably, the scientists interviewed were very much the author's personal selection and many individuals who are also making, or have made, important contributions to science have not been approached. Nonetheless, these eighty people do represent most of the chief strands of thought among scientists today, both in terms of their scientific interests and their attitudes towards those who determine the nation's science policy.

I approached the writing of this book as someone who has had the classical British arts-based education with no tradition of science in his family at all, except for a distant genealogical link with the 18th-century scientist, Henry Cavendish. Science appeared to me as a child dull as well as often being incomprehensible and based on a set of laws that could not be questioned. In contrast, the study of history, which became my main interest at university, was open to endless interpretation. It appeared dynamic and alive, with many of the issues that had precoccupied people in the past continuing to be relevant today. However, my developing interest in business, management, entrepreneurship and corporate strategy subsequently began to draw me into the world of science steadily and inexorably. For here I saw the impact of science and began to meet scientists. Some of those I met had been responsible for marvels of technology; yet no one seemed to have heard of them – often, being of a reticent disposition, they seemed to prefer to remain unknown.

When I began this book my aim was to talk to scientists about their work and their interests, as well as about the problems that I knew they faced in obtaining adequate funding. The issue of the two cultures was a secondary theme. However, the further I proceeded, the more it began to impinge upon our discussions and upon my thoughts about the scientific profession. As I talked to scientists and reflected on what they had said, it rapidly became clear why there is a gulf between the scientific community and those outside it. Firstly, most scientists – even those working in industry – spend most of their time working with or writing for other scientists. Theirs is an intense profession, demanding long hours, in which ideas and problems from the working day linger on into the evening and surface on weekends. As one scientist in industry said: 'This is not a 9 to 5 profession.' So scientists do not have the time to talk to non-scientists and their work does not encourage them to do so. When they do talk to non-scientists they realise very quickly how little science most of us know, which makes the task of trying to explain what they do almost impossible for them. Scientists find most of their friends, and often their spouses, too, among their scientific colleagues. So an important barrier to communication, which some scientists recognise, is one of language. The language of science is based on a multitude of definitions and concepts about which most non-scientists know extremely little.

Secondly, in this professional climate ideas and concepts are paramount and everyday prejudices are subjected to scrutiny. As another scientist said: 'Science teaches you to be sceptical and it encourages you always to ask why.' Huxley observed a hundred years ago that 'the great tragedy of science is the slaying of a beautiful hypothesis by an ugly fact.' Science is therefore always challenging our conceptions about how we think the world behaves and how we would like it be in the future. It is not a respecter of social hierarchies or tradition, which in part explains why a country like ours that is so wedded to both has an ambivalent feeling towards science and scientists. This ambivalence stems from distrust of its power, sitting alongside our reliance on its output. We want science to solve our problems but we do not want to feel beholden to it. Thus the second and much more important barrier to communication is the intellectual independence and rigour of science, which tends to set it apart from other careers like law, accountancy or business management. Within these professions, intellectual ideas have to be adapted to pragmatic human needs, otherwise their practitioners just could not function. But in science the rightness of ideas is what counts. As one physicist said: 'In science you can't be half right, you have to be completely right.'

Most scientists enter the profession because they enjoy discovering things. A number of scientists say that they do not just want to accept the world as it is given, they want to engage with nature, understand it and find out its secrets. They are therefore very curious people, weighing one fact against another in order to reach a higher truth. They are in a sense looking for a kind of certainty. So a belief without the support of coherent and testable facts is not attractive to them. This attitude is exemplified by Roger Penrose, Professor of Mathematics at Oxford and a cosmologist, who says that while he is interested in mysteries he is not interested in mysticism. He says that people of a mystical turn of mind have found comfort from his latest book *The Emperor's New Mind*, with its asymmetrical view of the universe in which time can run both backwards and forwards, but he prefers to maintain a distance from mystical interpretations of such a universe. It is the solving of mysteries that makes the game of science so appealing.

The process of discovery and the obvious enthusiasm that it generates is one of the themes of the book. Many scientists talk about their discoveries in a spirit of rapture, comparing the sensation they produced to falling in love. After years of painstaking experiments, which may have run for a decade, they suddenly see something that no-one else has ever observed before. Max Perutz describes the day he finally worked out the atomic structure of the haemoglobin molecule – after 22 years of study – as the most exciting day of his life. Although this is now thirty years ago, his face vividly lights up as he remembers it. Similarly, Jocelyn Bell recalls her first recording of a pulsar star with great emotion and this excitement was heightened when she discovered a second pulsar which, therefore, in scientific terms confirmed the first.

Indeed for many scientists, the confirmation of a discovery by oneself or others, is another source of pleasure. Most new scientific ideas have to fight quite hard for acceptance, especially if they are unexpected or if they confound existing theories. Here, personalities, entrenched prejudices and fashion play their part. The late Dr Peter Mitchell was derided for years because a novel theory that he had in biology regarding the way cells convert and conserve energy went against all the received thinking; it also cast a shadow over certain scientists' reputations. Denis Noble, Professor of Cardiovascular Physiology at Oxford, says that one of his theories was at first regarded as so implausible that no-one would publish his scientific papers. Both Mitchell and Noble were vindicated in the end both by subsequent tests of their own and by their critics' replicating their experiments.

The striking feature about these scientists is how their work builds on the work of others. Sometimes it is old ideas or discoveries long buried which trigger some new way of thinking, as when Professor Harold Hopkins recalled some of the experiments of the Victorian scientist, Tyndall, when he was trying to design an endoscope for surgeons. More often it is some relatively new discovery from within the last decade that stimulates a development. A case in point is Alan Devonshire's work on pesticide resistance which drew some of its inspiration from work by researchers studying cancer cells. Scientists are intellectual entrepreneurs who often to say to themselves, 'If that process works in that situation might it also work in this one?'

What these scientists are also saying is that a great deal of scientific discovery and invention happens by accident, which is cold comfort for the funders of science in industry and government. But it happens to be true. In the late twentieth-century world, where it is expected that we can manage everything, the accidental nature of science is not understood. It is certainly not accepted. Science works by asking questions, but it also works by following up by what may appear to be stupid answers and it works, too, by scientists going out on a limb. The stories in science often sound like detective stories in which one authority reaches an eminently sound conclusion which is accepted until another scientist casts doubt on some piece of evidence or suggests that another piece of evidence should never have been overlooked. This partly explains why scientists are somewhat wary of overly neat conclusions, because they worry that some telling piece of contrary evidence may have gone unnoticed.

So a major message of these scientists is that if industry or government restricts the scope of science to areas which appear to be the most worthy of support, either academically or industrially, then in effect we start to build walls round the subject which, in the long run, will limit and hamstring science. We will not get the ideas that we expect by restricting our focus. As Dr Peter Mitchell said, science has a habit of surprising you because it often gives you a completely new idea from a source that you were not expecting.

The real geniuses in science are those who can take various pieces of accidental discovery and create something new out of them. This is one of the most creative, exciting aspects of science – integrating information and discoveries into something wholly new. This integration of old ideas into a new concept often results in some tangible application. Yet when this happens the accidental nature of the original discoveries goes unnoticed. The advent of laser surgery for eye and tumour operations which occurred in the early 1980s was based on the invention of the laser which goes back to the 1940s. For years, the laser was regarded as an idea looking for an application. What made laser surgery possible was microelectronics and silicon-chip technology, both of which were inconceivable when the laser was first invented.

The near-impossibility of forecasting the results of science makes the choices involved in funding it extremely difficult. Increasingly, over the last five years, government funding bodies and commercial companies have tended to put much more money into science projects which appear to have a useful and predictable outcome rather than into more speculative ventures. If money is short then it is perhaps natural to put it where you expect the most obvious benefit. But this raises the question of whether we should see science as a cost or as an investment. Is science something that we spend money on grudgingly, a necessary but unwelcome expense, or does the money spent represent an investment with long-term benefits? The evidence in Britain is that we regard science as a cost which should be minimised as far as possible, while all our major competitors treat science as an investment. Although we do need to make choices in science funding, a major problem is that the pool of money available for implementing those choices is much too small, even for a country of our size.

A telling point made by many scientists – in industry and in academia – is that if we really do want to have 'useful science', i.e. science that gives us tomorrow's products and breakthroughs, then it can only come out of a strong science base. The leaps in science only come from a base that is being nourished. This means, in the industrial context, that companies need to maintain some research capacity of their own to enable them to seize a market opportunity when it comes, and, in the universities, that the whole system needs to be healthy otherwise even the centres of excellence will wither because their roots are not being nurtured. In neither case is the solution to buy in expertise from abroad at some later date, because to do so would be prohibitively expensive and also unlikely to be fruitful since we would no longer have the understanding to make the most of the imported expertise. Nor is it realistic to think that you can wait until scientists in other countries have published their results and then just copy them. One young scientist explains: It is a naive suggestion to think that you can stop your own research and just copy what other people have done. You would still need a research base. More to the point, what you tried to copy would be out-of-date by the time it was published.

There are two important issues in the funding of science. One is how we are to make the most of those scientists whom we have trained and how we can best exploit the new ideas of science. The second is how we should structure the funding institutions and bring some coherence between all the bodies involved. There is an awful wastage of talent draining out of science into other professions in the UK, or into other science centres overseas. All the scientists I talked to are worried by the brain drain. Government ministers may quibble over statistics and argue that the outflow is more or less balanced by the flow of people coming in. But to heads of department in our universities these figures are not mere statistics; they can see with their own eyes young people leaving and they can see the number of applications for posts – even in Oxford and Cambridge – diminishing. In any case, they add, most of those who come into Britain do so on a temporary basis while those who leave stay away forever.

There is also a waste among those who work as scientists in industry. Few companies make the most of their scientific talent, preferring to keep their scientists at arm's length rather than making them full and equal members of a managerial team. This is one reason why so many of the efforts to improve the transfer of technology from universities to industry have failed to live up to expectations; you can only successfully buy in and exploit technology from outside from a position of equality and mutuality. A company's managers and its technologists have to be able to talk more or less on equal terms with scientists outside if they are to work together. Moreover, the scientists and technologists need to have a status that is at least equal to that of the accountants. The head of Sony, Akito Morai, on a visit to Britain in 1991, said that he would never give accountants the kind of status that they enjoyed in this country. British industry still has not understood that accountants do not create wealth. Has anyone ever heard of an accountant coming up with a new product? But we can all think of at least a few scientists or technologists who have.

The second problem in the funding of science is that we have a number of institutions and advisory bodies which overlap each other, which do not communicate with each other sufficiently, if they communicate at all on occasion, and which do not have significant contacts with the users of science. If we had pots of money to throw around, this state of affairs would not matter, but our resources are unhappily severely limited. In addition there is an amazing amount of secrecy surrounding these bodies so that members of one scientific body will not know what members of another body are doing, even though they both deal with questions of science funding and policy. Side by side with this secrecy there are sudden shifts in policy on the part of politicians, which may render the work of these committees and bodies superfluous. If the advice given is so important that it has to be secret, why is the advice so often ignored or overruled by an ill thought-out government initiative?

What scientists frequently ask is this: Is there, in fact, a science policy for Britain? By this they mean not a policy for the next two years, but one that is intended to last for at least five, possibly ten. Along with shortage of funds, it is inconsistency that most bothers scientists and which is driving them out of the profession. Scientists need consistency to be able to plan their careers, to plan the scale of projects and to participate in international research projects of the kind that are too expensive for any one country to undertake. Industry also needs to understand the government's long-term policy on science, and not just in order to plan its budgets for research. Those companies which do use scientists in large numbers, those involved in chemicals and pharmaceuticals for example, need to know how many graduates are coming out of universities in four to five years' time.

Yet it is clear from talking to scientists and reading government and parliamentary reports of the last twenty years that we have had neither coherence nor consistency. Part of the problem is that science policy has become very short-term. We need to think ten years ahead and then provide adequate resources to sustain national objectives.

In order to make the most of our scientific resources, we need to reconstruct the machinery of science funding. As Sir Mark Richmond, Chairman of the Science and Engineering Research Council powerfully explains in the last chapter, we need to bring some coherence into policy-making. Sir Mark confirmed my growing view that we need a new body on the lines of the proposed Science Directorate outlined in the final chapter. This would bring together all the parties interested in science – from industry to the universities. It would be a statutory body like the BBC or the Arts Council, with most of its money coming from the government but some from commerce. It would also be a high-profile body, preferably headed by a chairman of independence and capable of articulating science to the public.

Science has always been the engine of material and economic progress, so in talking about science and debating the role that we want it to have in our lives we are also talking about our own future. It may be that we are disenchanted with science, that we feel that it has changed our lives too quickly and that it contains dangers. But if we are to distance ourselves from science and abstain from its fruits then we have to be aware of what we are doing. Science plays a far greater part in our lives than most of us imagine, influencing the food we eat, the drugs we are prescribed and all the electronicically-based features of our society. Our ability to pump North Sea oil – indeed our knowledge that oil existed in the first place – was all based on science, and on some very esoteric, seemingly useless science too.

In a country like Britain, which ranks among the highest in Europe in *per capita* ownership of computers, mobile phones and video recorders, science and scientists ought not to be remote. A major aim of this book is to make them much less so.

Part I
THE TRADITION OF BRITISH SCIENCE

— 1 —
WHAT MAKES A SCIENTIST?

The most beautiful experience we can have is the mysterious. It is the fundamental emotion which stands at the cradle of true art and true science. Whoever does not know it and can no longer wonder, no longer marvel, is as good as dead, and his eyes are dimmed.
Albert Einstein in *Ideas and Opinions*

Those comments of Einstein typify the true scientist's motivation: a sense of wonder coupled with a passionate desire to understand the source of that wonder. A scientist is like someone who hears a piece of music and not only wants to listen to the melodies but to read the musical score as well. It is that desire to know the structure behind appearances which distinguishes the man or woman of science.

Yet what is it that makes a scientist? What are the attractions of science and what do scientists mean when they talk about 'good science'? Why do some scientists spend all their lives in pursuit of the answer to some scientific question and never reach it, while others walk away with the prizes? How much luck is involved in the process of discovery and how far are the major advances of science due to dogged persistence or to inspiration? Science is both competitive and collaborative, but sometimes it is hard to tell which is the most dominant feature. Scientists naturally talk a lot about the scientific approach to problems, by which they usually mean objectivity; but at the same time there are strong emotional undercurrents in science, with prejudice and intuition often fighting out the battle of new ideas.

When scientists talk about science and the excitements of their work a number of words come up again and again in their descriptions of what they do. Words such as puzzles, crosswords, games, clues and mysteries pepper their way of looking at the world. These are people whose driving interest is to understand the world, who are fascinated by intellectual problems, who enjoy solving riddles. This relentless quest to solve often very difficult problems can lead to the pursuit of science almost as an obsession.

James Gowenlock, Professor of Chemistry at Heriot-Watt University, speaks for many scientists when he says that what drew him to science was

. . . the sheer intellectual fascination of it. In my case it is a love of molecules. You let them take over your imaginative faculties. You dream molecules. It's a love of intellectual problems. You find something that you don't understand and so you seek an answer, what is the pattern?

Sir John Vane, the Nobel prize-winning biologist, says that he has been driven by a sort of quest. 'Discovery is one of the excitements of life. It's enormously exciting to find something that's been there all the time but which nobody has seen before.' Both Vane and the French Nobel prizewinner, Jacques Monod, refer to the story of Michaelangelo, who was praised for a piece of sculpture, whereupon he apparently replied, 'But it was inside the stone all the time.' Science is exactly like that, says Vane. 'You're discovering what has always been there.'

A similar view comes from Sir Hans Kornberg, Professor of Biochemistry at Cambridge:

The satisfaction of science is seeing things in a context that was denied to you five seconds before and which transforms everything. You are making sense of something that had been confused and chaotic – it's like doing a crossword puzzle. I find research tremendously stimulating; it's an enormous satisfaction to do something which everyone says wasn't possible. When you do an experiment really well and you realise something that you never realised before, then you perceive a little tiny glimpse of the way the universe hangs together, which is a wonderful feeling.

It is this process of discovery that makes the practice of science 'a passionate activity', for Gowenlock and many other scientists who also talk of the strong emotions that they feel in the course of their work. To non-scientists science can appear to be coldly logical and lacking in feeling. But to those scientists, like Gowenlock, who are deeply engaged in their subject it is quite the opposite. As he says: 'To discover something in science is a living, creative activity because you are involved in understanding nature and the enjoyment that you have is the by-product.'

Some scientists, especially those in physics, astronomy and cosmology, find that science is for them a sort of 'quasi-religious quest'. This is how Paul Davies, a mathematical physicist, describes his passion for his own subject. He remembers as a child arguing with his curate about such things as the nature of consciousness, free will, the origin of the universe and the nature of time. The answers he received were unsatisfactory. Instead he turned to the works of popular scientists such as Sir James Jeans and Arthur Eddington. One of Jeans' books, *The Mysterious Universe*, sold so well that it was reprinted six times within two years' of its publication in 1930.

Jeans wrote with a vividness and simplicity that is rare for scientists, and reading his book today it is easy to see how he would have stimulated a budding scientist. Other influences upon Davies included the works of astronomer Sir Fred Hoyle. These scientists attracted Davies because they were 'grappling with some of the deep issues of existence'. So, from an early age, Davies says, he was 'convinced that the way to deal with these ultimate topics was through physics'. He adds that the same sort of quest inspires many other scientists.

Another attraction, which is frequently mentioned, is that science is international. It shares a common, boundary-breaking language, and its practice puts people of different nationalities in touch with each other. It is a much more universal subject than most others. Gather together an international group of particle physicists and they will all be talking about the same subject using the same descriptive terms; they will differ in what they regard as important, or in what they believe is the next step to pushing the subject further, but they will at least be in agreement when they use words like electron or proton or electromagnetism. By contrast an international assembly of historians debating a subject will come to it from very different perspectives, with the result that they will not share a recognisable contiguity of views about how to approach it. Historians from the USA or Germany or Africa will attach slightly different meanings to the words that they use and the importance that they should have. In science the language is common to scientists of every country, which is one reason why ideas in science flow quickly around the world.

Jean-Patrick Connerade, Professor of Physics at Imperial College, remembers coming face to face with this internationality on a visit to America. He recalls:

> When I was travelling in the USA some years ago I met a scientist and he knew immediately what I was doing which rather surprised me. He told me that there were only about 200 people in the world in my sector of physics and he knew about each of us. That international dimension is true of every part of science. I think you could drop a scientist into any major capital and they would know someone in that city in their activity.

Richard Hooley, a young biologist working in agricultural genetics, also alludes to the international dimension in science and its attractions. He says:

> Part of the satisfaction of working in science is mixing with other scientists and being part of a big scientific community. Meeting other scientists is one way in which you push forward knowledge; you discuss ideas and you give papers at conferences and you listen to other scientists. So you are part of a continuing dialogue in your subject.

Central to the international discussions that go on in science are the academic journals. Most scientific disciplines hinge on the publication of new research papers in journals devoted to specific branches of science. In molecular biology alone there are more than thirty leading journals that come out monthly or quarterly. These are required reading for research scientists, both in the universities and in industry. Leslie Iversen, Research Director at Merck Ltd (part of the US pharmaceuticals giant, Merck Inc.) has a weekly reading list of about twenty journals which sit on his desk to a depth of six inches.

These journals are important not only for science, but also for the careers of scientists. It is a principal way of receiving recognition. Reputations can be made with academic papers, and broken too. Even if scientists are not keen to publish there is strong pressure on them to do so. Promotion and funding partly depend on the publication of papers – the more respected the journal, the greater the kudos for the scientist. Career prospects have always been influenced by the quantity of academic publications, but in recent years scientists' published output has also become a key factor in whether they secure funding. In the 1990s most scientists are having to write their academic papers in their leisure time simply because they are too busy during their working hours.

The international dialogue between scientists is not only expressed in the publication of academic papers; it also involves actively working with other scientists on particular problems. For example, at the Long Ashton Research Station scientists from about six European countries are engaged on various agricultural studies and at the Monks Wood Experimental Station staff scientists are collaborating with partners in Europe on a number of environmental questions and are even bidding for consultancy and research assignments together.

Most scientists are stimulated by science at quite an early age. The child's chemistry set has often been father to the scientist. Vane was given a chemistry set when he was twelve, which he used in his mother's kitchen until one day one of his experiments blew up. His father had just painted the kitchen. Vane was subsequently given what was virtually his own laboratory when his father built him a shed in the garden fitted out with gas, electricity and water supply. He continued to do his own experiments there completely unsupervised. Family tradition has also sometimes been the spur to go into science. For example Professor Richard Gregory's father was an astronomer running the London University observatory, while Roger Penrose's father was a professor of genetics. In the first half of the twentieth century the most famous example of parental influence was probably Lawrence Bragg who followed his father to a Nobel prize; they are the only father and son to have achieved this distinction.

The most influential factor for most scientists was their school-teachers. Max Perutz says that it was the excellence of his Austrian chemistry teacher which made him want to take up science: 'I fell for chemistry at school because the teacher had organised practicals where we could do our own experiments.' When Perutz won the Nobel prize his teacher was one of the first to congratulate him. Perutz adds: 'It is a pity that science is so badly taught that children discuss science as boring. The wonderful thing about science is that it isn't just what you read in books:it is something that you do.' James Gowenlock tells a similar story of being influenced by his school's chemistry teacher, adding that from his class alone three pupils became professors.

What is even more revealing is that a number of these scientists said that they were almost turned off science when studying at university. When Vane was asked by a professor after he had finished his studies what he wanted to do next he replied, 'Anything except chemistry.' He was offered a job as an assistant to a professor of biology which he took with alacrity. Tom Blundell, Chairman of the Agriculture and Food Research Council, spends part of each morning in his laboratory if he can possibly manage it, so keen is he to maintain an active interest in research. Yet he says that he only appreciated 'the excitement of science' once he started to do postgraduate work as a member of Dorothy Hodgkin's team researching into insulin. He says: 'It was only then that I realised that science was about being critical and imaginative and international.'

These experiences recorded by scientists raise the question: what makes a good scientist? One thing that can make a good scientist is having a first-class supervisor in their early career. Vane says that working in Oxford under Harold Burn in the 1950s, turned him from being an amateur into being a professional.

> He taught me the importance of reading the history of my subject, reading the literature before I made experiments and then showed me how to conduct experiments in a rigorous way. He really enthused me and showed me how to think scientifically.

James Gowenlock remembers beginning his career under Michael Polanyi, whose father was a Nobel laureate. Polanyi, he says, was a great influence upon him, 'always talking chemistry, provoking questions and challenging me to think harder'.

Patronage can also be an important influence. Many of the scientists in this book have worked early on in their careers for a distinguished professor who has helped to open doors for them. An example is Sir Hans Kornberg who left school at seventeen to work as a laboratory technician under Hans Krebs at Sheffield University. It was Krebs who persuaded him to take a chemistry degree and who got him started as a biochemist. Patronage of a kind also works in mid-career, and for some scientists has been crucial in securing an appointment. This was especially true during the early 1980s when senior posts were cut back. These personal influences operate across national boundaries, with British professors recommending clever pupils for two-or-three year research contracts in American research centres where they have contacts, and vice versa.

Most scientists, in talking about their work, at some stage use the phrase 'good science'. This phrase comes across in conversation and in articles where they argue for more funding. They are often to be heard saying that if we do not fund science sufficiently then we will not get good science; or that if the government funding agencies tell scientists what research they should be doing we will not get good science either.

So what do they mean by 'good science'? Sir Hans Kornberg:

> Good science is the ability to look at things in a new way and achieve an understanding that you didn't have before, it means looking at fundamental processes and having the ability to ask questions so that you can formulate the kind of strategy that takes you to the next stage. It is opening windows on the world.

Annette Dolphin, Professor of Pharmacology at the Royal Free Hospital, London, believes that for a piece of work to be described as 'good science', 'It has to be novel – there's a lot of me-too in science – it has to be published in a good journal and it has to be widely quoted, so that it sparks off other people to do more experiments.' In other words, it has to be a little out of the ordinary. She elaborates:

> One could do something very well and publish it, but if no one ever read it or included it in their own ideas I suppose one might still say that it was good research but it would mean that it was a dead end that you had been working on, that is, something that no one was interested in.

However, this assumes that the view of one's peers in science are always right. Very often they are; but in at least 10% of cases a scientist's peers are wrong. It can sometimes take a scientist five or ten years for his or her ideas to be accepted by the scientific community. Professors Denis Noble and Colin Blakemore and the late Dr Peter Mitchell recall some of their ideas being greeted with suspicion and having to wait up to ten years before they were completely vindicated.

Yet there is more to good science than having an original view. For Roger Penrose, there are two additional aspects:

> There are two features of all good science. One is having broad goals and the other is just enjoying your area of science for its own sake. To make real progress towards the big goals you just have to enjoy playing around with the subject. So in my case I've been intrigued why the laws of physics don't distinguish between the past and the present, and that has led me to the broader question about the initial state of the universe in order to explain those laws.

These perceptions of science beg the question of how scientists can know whether they are wasting their time. It is an important point, because scientists frequently say to politicians that scientists are the best judge of what they should and should not research. Sir Hans Kornberg says that a good scientist will always know when he has done good work: 'He knows it in himself. He is the one who knows whether he has been creative or not and whether he has really asked fundamental questions.' Professor Dolphin agrees:

> You always know if something that you have done is good or not. I will know if something I have done is good or not because I will know that certain things have pushed my understanding forward of what is going on, that I have made a quantum jump in some way and that I've found something that couldn't have been predicted.

Many people, including scientists, ask themselves what the factors are that bring scientific success. Why do some people make the outstanding breakthroughs when others do not? The late Dr Peter Mitchell recalled hosting a conversation in his home in 1990 at which he discussed this with five other scientists, all of whom were, like himself, Nobel laureates. The other five were Frederick Sanger, Lord

Porter, Maurice Wilkins, Sir John Kendrew and Herbert Simon. As he re-collected the occasion:

> We sat and asked, how had we done it and none of us had any idea at all. We all had different minds. Kendrew had been very methodical and imaginative, Porter had been logical but didn't think he had a very classificatory mind and so on. Maybe it was imaginativeness. We had nothing in common except perhaps an uncanny imagination that enabled us to say I think that this problem can be solved and I've got the faith to to go on working at it for ten years or more.

Mitchell added that there are really two kinds of scientific discovery: the technical discovery, such as finding the structure of a molecule, and the theoretical discovery like Einstein's theory of relativity. The first reveals something that is known in greater detail whereas the second covers general principles. However, he also pointed out that general principles may result from a technical discovery, citing the molecular work of Perutz and Kendrew on the haemoglobin and myoglobin molecules.

Is the making of a discovery perhaps simply a matter of having sufficient intelligence – is the crux of the matter that some scientists are just cleverer than others? Again, Dr Mitchell answered this with a story. He recalled a discussion in which the conversation turned to another scientist who had recently been awarded the Nobel prize. One scientist expressed surprise that the prize had been given to someone who had never appeared particularly intelligent, but another scientist remarked that this was the whole point, saying, 'If he had been intelligent he would have given up the whole project long ago, because he would have deduced that it was insoluble.' Mitchell commented: 'That was a very profound remark. A lot of good science is the result of very good problem-solving rather than high intelligence.'

A similar point is made by George Efstathiou, Professor of Astronomy at Oxford, who says:

> If you look at those scientists who are Fellows of the Royal Society, a higher than expected fraction do not have first class degrees, but they have all notched up achievements in science. There is a certain threshold of ability, of understanding, that you need to pursue science. But beyond that threshold it is more important to be energetic, enthusiastic and original and you can make up some deficits in ability by persistence and application. Application is crucial. One thing that the public doesn't understand is how incredibly hard most scientists work.

Many scientists talk about luck, of having the right idea at the right time, or of coming across an article that triggered a new experiment leading to the solution of a problem. So how important is luck in the scientific process? Cesar Milstein, who discovered monoclonal antibodies, says: 'There's an enormous amount of luck but also an enormous amount of intuition.' Others, such as Annette Dolphin, think that luck can be important but that this element always has to be seen in context:

To some extent there is an element of luck, but it's not really luck because you can look at some scientists and you can see that that they've done novel things in their career twenty times in their lives. Certain scientists have this knack for discovery. It's not so much luck as in knowing what areas to pursue. A scientist might ignore something because it wasn't expected and rationalise what's happened by saying that the experiment went badly, whereas another scientist might decide to follow up what appears to be an odd result.

Noticing an unexpected quirk in an experiment and following it up explains why, for example, pulsar stars were discovered in Britain and not in America. Anthony Hewish and Jocelyn Bell were scanning the sky for quasars and noticed that their recording equipment was charting an unusual phenomenon. They followed this up, and their subsequent monitoring of new recordings led to the discovery of pulsars. Meanwhile an American team also noticed some strange readings on their recording equipment. They thought their machine had made a mistake so they knocked it as one would knock a television set to improve the picture and went home. In any case, the American team was working on another project which was running late so they had no time to enquire too deeply why their machine was making the occasional odd recording.

In fact, it is often the case that a scientist will only notice what he or she is looking for. This necessary selection of scientific data for analysis has inevitable hazards, especially if the selection is programmed by computer. A case in point is the discovery of the hole in the ozone layer by a British scientist, Joe Farman. When the discovery was published, the initial response by US scientists was that the finding must be wrong because their own computers – analysing atmospheric observations – had not picked this up. The reason was that their computers had been programmed to look for other, very specific phenomena, and so the ozone hole was screened out of their analyses.

Knowing how to choose the next step in research is important. Kornberg recalls a comment from his early mentor, Hans Krebs, who told Kornberg that 'A good biochemist knows what experiments he should be doing but a first-rate biochemist knows what experiments he shouldn't be doing.' This is a point underlined by Dr Cesar Milstein: 'You can attack a problem in a thousand different ways. What is important is the questions that you are asking and how you go about asking them.'

Most scientists say that it is impossible to plan a piece of research every step of the way. If the study in question were to be so predictable, then that would presuppose that the answer was in part obvious and therefore the question would hardly be worth asking. In addition a scientist always finds new things to look at as the research unfolds. Michael O'Shea, Professor of Biology at Sussex University, believes that in fundamental research it is difficult to plan a project for more than three months. This goes flatly against what the government funding agencies would like to hear, as O'Shea readily admits:

It is very hard to plan research in the way that the Research Councils would like to see you do it. Of course you write a plan. But part of the skill in research is detecting where to go in an on-going project where the goal you set initially may appear less interesting than when you set it.

While having the imagination to follow up something unusual is important, what is also necessary is to be aware of the work of others, both those in the past and in the present. Dr Peter Mitchell explained: 'You're absolutely dependent on the work of other people. Discoveries are not really unique to one person.' Colleagues play a part – even if they are not directly involved – as Francis Crick pointed out in his own account of the discovery of the structure of DNA. So, too, do scientists working in other centres, which again is why international conferences and academic papers play such an important role in the process of science.

Being familiar with the experimental techniques of others is also significant, says Jean Thomas, Professor of Macromolecular Biochemistry at Cambridge:

A molecular biologist, for example, needs to understand the biochemistry and the biology of a system they are studying which means that they need to know about the work of people in those areas, how they have carried out their experiments and how they have reached their conclusions. They might want to know the structure of a particular compound as a crystal but that means that they first have to have an appreciation of the capabilities of the techniques that they want to use, i.e. what state should the compound be in an experiment, how they should cut it into pieces and then into what sort of pieces. They need to know what they can get out of an experiment, but that depends on what sort of image resolution they want at the end. Everyone is building on the work of others.

This comment also partly explains why scientists emphasise that, as a nation, we cannot simply leave research to other countries and then come in afterwards and pick up the conclusions – unless we have done similar experiments to our competitors we will not understand how they have arrived at their results.

Science has become increasingly competitive since 1945. More countries recognise the economic and military value of science and have put an increasing amount of funds into it. There are also more scientists. The competition to be the first with a new discovery or to make a breakthrough in a technique has become intense. This partly explains why science has splintered into ever-narrower disciplines; one way to make a name for yourself is to carve out a new area that hardly anyone else has thought of or to join the select band of those who have marked out a new territory. The competitiveness of science is something that nearly every scientist talks about. It can determine their choice of post-doctoral study, which professors they seek to work with, which universities they search out, etc. It undoubtedly shapes their careers to the extent that they may forsake their favoured subject in order to shine in another.

Yet, to outsiders, this competitiveness seems to sit oddly with the urge to publish as soon as you have achieved something new. Why give your competitors, especially those in competing economies, the benefit of your latest results? Scientists say that it is not as simple as that. What they publish rarely has an immediate commercial application, while some scientists add that what they are doing when they publish research results is to state a position without giving away all the details of how they have arrived there. In fact, it is the very competitiveness of science that spurs the publication of research results.

Increasingly sophisticated equipment plays an ever greater part in scientific research. It is no longer enough just to have good ideas. What is equally important is having the technology to prosecute those ideas well. As Professor Jean Thomas says:

> Without the best equipment you can't push your research to the limits. In my field if you want to understand molecular structures in detail you have to look at them at the atomic level which can only come from X-ray diffraction or magnetic resonance spectroscopy machines which are expensive.

The technology of new equipment is also speeding up scientific advance, as Annette Dolphin explains:

> In my own area a piece of equipment called a patch clamp has transformed research. This equipment is based on a technique which allows you to record from single living cells their electrical activity. With that equipment you can see a single ion channel in the cell opening and closing. So technical advances enable you to do things that you couldn't have done before at all.

A less happy consequence of this rapid technical development is that equipment becomes obsolete very quickly. So the advances in technical equipment are both a boon and a curse for science. The new equipment offers untold opportunities to the scientists but it also makes the practice of science more competitive because part of the secret in pushing science forward now is to be found in the ability of the scientists to operate new and sophisticated instruments. In addition the new technology raises the cost threshold for research. New equipment is expensive and it has to be serviced and managed by technicians. Technical support now plays a much greater part in scientific departments than it did twenty years ago.

The environment of science, the questions that it poses and the methods that it uses, appeal to a certain kind of mind. Leslie Iversen, who has spent a lifetime working with, recruiting and managing scientists, has one view on what the scientific mind and personality is. He says: 'I think science appeals to people who are easily bored, who are restless and who don't like routine.' But to the non-scientist, and even to many scientists themselves, the experimental basis of science is highly routine. Iversen disagrees:

> Experiments are not routine in basic science. In basic science you never know what

you're going to be doing in six months time; the field will move on and you will move with the field and you will look back and say how ignorant I was six months ago. He adds: 'A very useful ability is to be able to see before others which way the field is moving and to make the decision to be in there.' Science also requires a particular kind of personality. 'You must have a genuine intellectual excitement in what you're doing. . . .It tends to be a bit obsessive.'

The all-embracing quality of being a scientist rings bells with Professor Denis Noble. He says:

> For many of us it becomes almost a consuming passion. So you find that there is a tension between your emotional life and your science and for a scientist these things compete. Science and family life compete for the emotional inner heart of a scientist.

The practice of science eventually shapes the way a scientist looks at the whole world and how he or she communicates with other people. Miles Padgett, a young physicist working with PA Consulting Group, says:

> Most scientists, especially physicists, tend to be very sceptical. A good scientist will always ask why. We're quite argumentative. Scientists do have enquiring minds, but it can make them appear very negative at times: it's much easier to disprove something than it is to prove something, because you only need one thing to be wrong with an idea to be able to cast doubt on it. The question a scientist will always ask, is not how can I help you, but why; that's the way science works, the way you're taught to do problems. You're sifting information in your mind all the time.

This inquisitorial, sceptical aspect lies at the heart of science. Its repeated questioning of received ideas and mysterious phenomena is what has helped science to advance, but at the same time, its challenging, almost iconoclastic, function has been a prime source of tension between science and the intellectual and social cultures with which it co-exists. In Britain this tension has produced a veritable split. Exactly why the British should feel the need to keep a distance from science more than most other countries is still a puzzle. The next chapter offers some further reflections.

2

REFLECTIONS ON
THE TWO CULTURES

*I believe the intellectual life of the whole of western society is
increasingly being split into two polar groups. . . . At one
pole we have the literary intellectuals (who incidentally, took
to referring to themselves as 'intellectuals' as though there
were no others) and at the other we have the scientists, and as
the most representative, the physical scientists. Between the
two a gulf of mutual incomprehension – sometimes hostility
and dislike – but most of all lack of understanding. They
have a curious, distorted image of each other. Their attitudes
are so different that, even on the level of emotion, they
cannot find much common ground.*
C.P. Snow: 'The Two Cultures and the Scientific
Revolution' (The Cambridge Rede Lecture,
Summer 1959).

Snow's commentary on the place of the scientific culture in society was directed at
all western countries, but his remarks were debated most fervently in Britain,
arousing a national controversy that went on vigorously for four years. This was
partly because one of his subsidiary themes was that it was high time that non-
scientists acquired a little interest in science. Snow himself was both a scientist
and a novelist and had worked in the civil service. The irony of his remarks was
that in 1959 science was at a high point in public esteem. Science appeared to be
capable of solving all our immediate material problems, and quickly too, and its
technological fruits were amazing the country. The Concorde aeroplane, for
example, was then in its second year of development. Yet alongside this acceptance of bright new technologies there was, as Snow rightly said, a growing gulf
between scientists and non-scientists and a suspicion of science. Nowhere was this
gulf greater than in Britain, the country where 25 years previously scientists had
succeeded in splitting the atom.

The cultural position of science in Britain has always been an odd one.
Although there is far more coverage of science today than there has ever been and
although one of the biggest bestsellers of recent years has been Stephen Hawking's

A Brief History of Time, science and scientists are still marginalised in our society. As a number of surveys have shown, public knowledge about even the rudiments of science is abysmal, with roughly one in three people in Britain believing that the sun moves round the Earth. In many of Britain's major institutions scientists are still kept at arm's length – not one Permanent Secretary in the civil service is a scientist – while in industry and the City scientists are regarded with suspicion.

Most foreign visitors look upon our attitude towards science as extremely strange. In Paris and Tokyo there are new exhibition centres for science which have proved immensely popular. In Britain's capital we do have a showcase for science which, to be fair, does attract lots of visitors but it is curiously called a 'museum'. The same group of buildings in which we house scientific artefacts also contain dinosaurs. These amusing absurdities almost confirm a notion that science is not part of the here and now. It is a genie in a bottle, to be brought out on special occasions and firmly put back when we no longer need it.

The separate functioning of the artistic and the scientific cultures have become so deeply rooted that it is difficult now to imagine a country in which the two were once part of a wider, unifying intellectual culture. The historical point at which the two cultures started to grow rapidly apart appears to be around 1870, though some of the root causes of the schism lie much further back. What is perhaps surprising is that the period that preceded this sundering of the two cultures was a time when science and technology caught the public imagination and it was also a time when a real effort was made to put science at the centre of British culture, as Professor John Durant explains:

> Victorian culture was absolutely bedazzled by science and technology in a way that would be hard to see today. The machine and the power of the machine were a dominant cultural motif, and the notion that, through science, economic and social and even moral progress could come about was almost a *leitmotif* of the Victorian period. In 1851 there was the Great Exhibition which illustrated the tremendous flowering of British science and technical progress. At the same time there was a bid by scientists such as Huxley, Tyndall and Spencer to give science a central place in British culture, and they were encouraged by a lot of intellectual and literary people, too, such as George Eliot and Sir Leslie Stephen. For a time it looked as if we might create a scientifically oriented intellectual culture in this country, but it didn't happen.

The reason it did not happen was the arrival of Darwin's book, *The Origin of Species*, in 1859, which caused a ferment of discussion among literary and scientific intellectuals. One irony was that Darwin's thesis on evolution was too readable. It was, says Durant, written in such a style that it was accessible to a very wide audience. But the success of Darwin's book was to break the common culture that had made its huge sales possible. According to Durant:

> You begin to see this common culture breaking down as people tried to forge a new

synthesis which will be consistent with the new science that Darwin's contemporaries were putting forward and compatible with the consensus of the religious and spiritually oriented.

Some of the Darwinian debate took place in a club called The Metaphysical Society which argued over the ideas inherent in evolutionary theory throughout the 1860s. Its members included many of the leading social, political and scientific lights of the day such as Gladstone and Huxley. They were looking for a new framework unifying science, metaphysics, morals and even the spiritual, but by about 1870 they admitted defeat. They realised, says Durant,

> That there wasn't a clear foundation for a common discourse any longer, with Gladstone concluding by saying that science should be left to scientists and theology and religion to philosophers and theologians.

The importance of Darwin's work in cultural and religious terms is underlined by Peter Mathias, Professor of History and Master of Downing College, Cambridge. He describes the evolution debate as a turning point for British science. What he finds interesting about this debate is that a century earlier clerics were among those advancing science not attacking it.

> In the eighteenth century most of botany was in the hands of Anglican clergymen. The natural world was seen as the revelation of God. When clergymen looked down their microscopes they were confident that they were seeing God's wondrous world.

However, like most arguments, the debate that followed Darwin's book was about several issues, not just religion. Most Christians, after all, eventually accommodated themselves to Darwin. Woven into the debate were issues of social progress and materialism and how much, if any, primacy should be given to science, technology and industry, on the one hand, and literature, ethics and artistic experience on the other. These apparent dichotomies were expressed by Thomas Arnold, headmaster of Rugby public school who said:

> Rather than have science the principal thing in my son's mind, I would gladly have him think the sun went round the earth, and that the stars were so many spangles set in the bright blue firmament. Surely the one thing needed for a Christian and Englishman to study is a Christian and moral philosophy.

Arnold's strictures against science were extremely influential, for his displacement of science in his own school was imitated by the majority of his public school competitors. Henceforth an educated man was one who had mastered the classics, philosophy and literature. Schools for women were also affected by Arnold's views; a government education review by the Taunton Commission in the 1860s found to its surprise that three-quarters of schoolgirls learned science but only one in five learnt Latin: the Commission quickly remedied what they believed to be a gross aberration by recommending that girls' schools follow the new educational norms of the boys' schools.

Science now became a vocational subject only: pupils only studied it if they were going to practise it, and since it was no longer considered socially acceptable to practise science there were fewer opportunities to learn it. Other subjects such as history or literature were fine – even though most of those learning these subjects would never expect to be historians or writers – because they represented true culture. The paradox of this attitude was that the social elite who were put through this kind of education, while believing themselves to be supremely pragmatic, were in practice completely impractical. But that was the whole point. The reason for the rejection of science, as a set of ideas and as a socially acceptable pursuit, was that it fitted in neatly with the rigid Victorian class system. Why bother with science and technology if the lower orders are going to be the technologists?

This is precisely what happened in the building of the empire: classically educated administrators ran the territories but if a railway had to be built then mechanics were shipped out to construct it and dispatched home when it was completed. The genie went back into the bottle again.

The tension between science and the ruling establishment was evident in the eighteenth century, when scientists like Henry Cavendish who were part of the aristocracy were frowned on by their own class. But the conflict was not absolute. Even then it was recognised that science had a military and naval value and therefore could not be totally written off. A century earlier Newton had underpinned British naval power by establishing the principles of longitude which enabled the admirals and merchant venturers to pinpoint the position of their ships. Some scientists believe that it was Newton's navigational work that helped to raise the position of science in seventeenth- and early eighteenth-century Britain and which ensured Newton's heroic and hallowed place in popular esteem. Jean-Patrick Connerade, Professor of Physics at Imperial College, London:

> The ruling establishment was very much in tune with Newton and he with them. One key reason is that he solved a very practical and important problem for them which gave their navy an edge over the navies of their rivals.

What subsequently made the conflict between science and the ruling establishment explicit was, firstly, Darwin and, secondly, the maturing industrial revolution. Darwin challenged the idea of a settled hierarchical order, and the industrial revolution with its attendant manufacturing magnates threatened it too. The scientific community from which Darwin came was weakened by the deliberate exclusion of science from public school curricula, while the rising industrialists were weaned away from their interests in business with titles and ennoblement. The process by which the leaders of manufacturing industry were seduced into the aristocracy and the gentry, and then sent their sons into the classics-teaching public schools has been ably described by Martin Weiner in *English Culture and the Decline of the Industrial Spirit*. Weiner describes Victorian industrialists as 'aspiring gentry', whose yearnings resulted in a haemorrhage of talented people from industry into the 'more congenial role of *rentier* country gentlemen'. The assault upon industrial values fed through into an assault on science, which had nourished the industrial revolution and expanded alongside it. So science was not so much attacked for itself as for what it led to.

There was however one group of people – the literary and artistic leaders – who did assail science for what it represented in intellectual terms. Their indictment of cold science and industrialisation provided a further powerful impetus towards splitting science from mainstream culture. One of them was Thomas Arnold's son, Matthew, who defined culture as 'the acquainting ourselves with the best that has been known and said in the world', and as 'the passion for sweetness and light, and what is more the passion for making them prevail'. For him science now lay outside culture. Others such as Ruskin and Rosetti sought to popularise this view, harking back to a golden medieval age where men and women expressed their creative natures by practising simple crafts in a pastoral world. Their perspectives were taken up by the educationists and provided the backdrop for what became the British liberal arts education. Thus social prejudice against science and industry was given an intellectual respectability which, in the long run, has been the most potent force ranged against science.

One early and little-known casualty was the division of the South Kensington Museum – inspired by Prince Albert – into an arts and science section. It was originally devoted to both 'arts and manufactures', and it was, according to John Durant, 'part of Albert's vision to embrace them both'. In the last quarter of the nineteenth century, the collection of objects was split, with the arts displays remaining in what was re-named the Victoria and Albert Museum and the scientific pieces housed in a newly-designed Science Museum based on the architecture of 11th century Romanesque France, with some references to the buildings of the collapsing Ottoman empire. It was clear, says Durant, that

> . . .the Victorians felt uncomfortable having the works of science exhibited with those from the arts, and the decision to split them shows just how far the two-cultures mentality was developing.

Yet science did not retreat from its role as a central activity without putting up a fight. On the one hand there were the scientists such as Huxley who argued that science provided as noble an insight into the natural world as any of the writings from the classical world. On the other hand there were also industrialists and scientists who argued that not only should science be respected for its own sake, but that it should be directed towards industry. Peter Mathias says that one of the things that has impressed him about the debate on science in the nineteenth century was

> . . .the discussion of science for its utilitarian purposes and the references to the absence of institutions for science and the absence of public funding of those institutions which would produce a progressive science base leading into industrial innovation and new ranges of industry, with Germany being cited as the main alternative model.

Many technical institutions, mainly in the Midlands and the North of England were set up as a result of this debate; these institutions were associated with particular industries, e.g. the Institute of Brewing. So there was a body of opinion that was keenly aware of Britain's declining technical capacity and which did succeed in creating some new institutions, but these institutions lay outside the mainstream educational system and they were not able to reverse the weakening of Britain's technical base. Imports continued to rise in those areas of industry associated with the advance of science, most notably chemicals.

One of the great puzzles of British science is that if the existence of two divergent cultures has been so damaging as this story suggests, why is it that British science has been so spectacularly successful in the last hundred and fifty years? There are two cardinal reasons. In the view of Durant, the first is the two culture system itself. The two cultures has produced

> . . .an extremely specialised form of education with those taking science receiving a very high quality education in their subjects, so that universities take people who already have a good grounding in science at a fairly advanced level.

To take an example, most academics would say that a British A level student in physics is probably on the same level as a first-year physics undergraduate at many a university abroad.

The second reason is the impact that war and the threat of war has had on British science. When the diplomacy of Britain's establishment fails, the country often turns to its scientists to get it out of difficulty. It is no accident that two of Britain's five research councils, which fund major areas of science, owe their origin to the First World War. They are the Medical Research Council, founded in 1913 as the Medical Research Committee, and the Science and Engineering Research Council which began as the Department of Science and Industrial Research (DSIR) in 1916. These two bodies grew out of the recognition that the country's physical and industrial health had slipped and needed to be improved quickly. Both of them developed quite rapidly, with the DSIR working closely after the war with a number of industries such as aircraft, engineering and chemicals. Looking back, says Mathias, it was remarkable how rapidly we caught up even by 1918. On the other hand the interest that British governments have taken in science for military purposes has had its disadvantages. Huge sums have gone into military research and weapon development, which has meant less funding for civil research and development. There have been some scientific trade-offs between military and civil science, but commercially they have been poorly exploited.

The Second World War was a major turning point for Britain science, for it led to two contradictory views about science. On the one hand it initiated a euphoria about the possibilities of science, especially its applications in the energy and aircraft industries. On the other hand the war's destructiveness brought an uneasy feeling; few people have described this better than the scientist, Jacob Bronowski, who visited Nagasaki on a warm November evening three months after it was destroyed by the second atomic bomb. His first reaction was that he

> . . .had blundered into this desolate landscape as instantly as one might wake among
> the craters of the moon. . . . What I thought to be broken rocks was a concrete
> power house with its roof punched in. The shadows behind me were the skeletons of
> the Mitsubishi factory building, pushed backwards and sideways as if by a giant
> hand. . . . What I met, almost as abruptly, was the experience of mankind. We
> looked up and saw the power of which we had been proud loom over us like the ruins
> of Nagasaki. . . . Conscience, in revenge, for an instant became immediate to us.

But Bronowski's response was to say that whether science is

> . . .dream or nightmare, we have to live our experience as it is, and we have to live it
> awake. We live in a world which is penetrated through and through by science, and
> which is both whole and real. We canot turn it into a game simply by taking sides.
> (*Science and Human Values* 1961).

In the early post-war years attitudes towards science continued to be a mix of
optimism and pessimism, but these competing stances were soon to be joined by
another, more insistent, feeling. This was that science was changing our whole
way of life – and not for the better. A year after Snow delivered his Two Cultures
lecture, Bronowski was writing:

> It is standard good form nowadays in polite society to say that the civilised values are
> disappearing, and then to blame science for the change. This latent opposition to
> science appears now whenever values are debated.

Those who complain about science, he continued,

> . . .feel that scientists have no spiritual urges and no human scruples, because the
> only success that science acknowledges is success in conforming to the material facts
> of the world. ('The Values of Science' – *The Rationalist Annual*, 1960).

To a large degree these reservations about science still exist, in part for the very
good reason that there are still two cultures. An additional reason for antipathy to
science has been the rise in environmental pollution: the dangers of the misuse of
chemicals, for example, or the problems of nuclear power, have become all too
apparent.

Some scientists feel that the distinction between the two cultures an artificial one,
and that they are really one because they both spring from human creativity and
aspirations. Max Perutz, for example, says:

> Science is part of culture. Culture isn't only art and music and literature, it's also
> understanding what the world is made of and how it functions. People should know
> something about stars, matter and chemistry. People often say that they don't like
> chemistry but we deal with chemistry all the time. People don't know what heat is,
> they hardly know what water is!

Perutz is dismayed by how little people know about science, adding, 'I'm always
surprised how little people know about anything. I'm puzzled by it.'

Other scientists, such as Richard Gregory, Professor of Neuropsychology at Bristol University, are not at all surprised at the depth of the divide between science and the rest of society. He says:

> I think it's quite obvious. Doing science is very different from doing the arts. Science is difficult. You need mathematics and statistics, which is dull like learning a language. The difference is that with a human language each word you learn you can use. In science you can work away for months and nothing happens. The arts intrinsically appeal to the human soul but a lot of science doesn't, and a lot of science is incredibly boring to do.
>
> Science is also an experimental philosophy, and philosophy isn't popular in Britain. The British people are not comfortable with anything intellectual or with things that are abstract and a lot of science is based on very abstract thinking.

One of Gregory's former pupils, Colin Blakemore, Professor of Physiology at Oxford University, argues that the reason why science has become unpopular in Britain is because the scientists themselves are so unknown. It is an extension of John Durant's argument that scientists have become invisible. The effect of this invisibility, says Blakemore, is that the public fears scientists because, while it does not know them or understand what they are doing, it is nevertheless aware of their influence. So scientists are seen as both powerful and alien, and therefore threatening.

The strangeness of scientists as perceived by the public is another factor. As Durant says:

> We are not well-served by the cultural stereotypes of the scientist. In our culture the scientist is either the mad inventor or genius like Einstein, who is always thinking incomprehensible thoughts, or the Frankenstein acting irresponsibly, meddling with nature and producing horrific things. The common factor is that the scientist is seen to be odd and peculiar. So people feel that they have no relationship with scientists and that is not true of other professions.

The uneasy relationship between scientists and non-scientists in Britain can be experienced in any large meeting that attempts to bring them together. The scientists will argue very rationally about the contributions that they have made to our society, comparing conditions today with those at the beginning of this century. They will also argue that without more science the social and industrial circumstances of our society will not only not improve but may in fact deteriorate. Non-scientists, sensing the power of science implied in such claims, then accuse scientists of being responsible for pollution and for the side-effects of modern technology. In addition there are often passionate complaints that science has robbed man of his soul. The two sides rarely seek to find any common ground, but instead argue from differing standpoints which are mutually incomprehensible.

What bedevils any discussion about the role of science, either culturally or politically, is that, rather than recognise that there are two ways of looking at the world, scientists and non-scientists often try to convince each other there is only one way. Each uses a different set of concepts to describe things, and has a different set of criteria by which to make judgements. Their separate ways of perceiving and responding to the world are not only the products of their education but also of the careers that they pursue. The practice of science shapes the mind, as does that of politics, accountancy or law; but in the case of science there are some very important intellectual differences which distance scientists from the public at large far more than is the case with other professions.

One of the surprising characteristics of many scientists is that they believe that their way of thinking is not very different from anybody else's. Many scientists say that what they do is just like any other intellectual activity. Lewis Wolpert, Professor of Biology as Applied to Medicine, writing in 1988 in *A Passion for Science*, says:

> Is there anything more to successful science than common sense, and the pursuit of logical internal consistency and correspondence with the external world? My own view is that what I do really differs very little in essence from the work of a historian; a search for explanation and connections, the process of validation or verification, the falsification of ideas. What makes the study of history different is less the approach than the subject matter.

That description of what a scientist thinks historians do is not one that historians would completely recognise. Of course both history and science attempt to arrive at explanations, and they both strive towards conclusions that have an 'internal consistency'. But there remain deep differences in approach. Few historians describe their published works as having a scientific basis, in the way that scientists use that term. Those historians who, like Marx, have tried to arrive at a 'scientific view' of history have generally been much criticised by their own profession. When a scientist and a historian talk about 'facts', they have a different image in their mind of what they mean. To a scientist, a fact is usually something that he or she has either tested or knows to have been tested. To a historian, a so-called fact will often be incomplete and will, by its very nature, not be susceptible to being tested in a rigorous sense. In addition, the 'facts' of history are often the least important part of it, with intentionality, beliefs and values – including those of the historian – being more important. Both science and history deal in hypotheses, but whereas in science a new hypothesis is usually based on new facts, in history the essential data may remain the same but the hypothesis may completely alter. The intellectual approach of science is quite special, and it becomes so much a part of scientists' way of thinking that they underestimate the extent to which their way of thinking differs from those of others.

One major difference is the precision of science, which is expressed in its methods and its language. Roger Cashmore, Professor of Physics at Oxford rightly says that in science it is not good enough to be half right, you have to be completely right. This striving for exactness and completeness is reflected in its language. Professor Gowenlock makes the point that:

The language of science, such as chemistry, has to be very precise because what you say has to have a precise meaning. As a university teacher you have to show your students how to use this language correctly because that's the key to their understanding.

Most other professions also use a language which is peculiarly their own, as anyone who has ever overheard a group of lawyers talking will agree. But the difference is that the languages of other professions use terms with which lay people are at least superficially familiar and they are continually being adapted by their links with the people whom they serve. By contrast the language of science is a closed system, which is necessarily bound to be the case. However, that does not mean that science as a whole is impervious to the values of the society in which it lives. As many scientists have pointed out, modern science is a product of liberal values, but the actual day to day language of science is a closed artefact.

Yet it is not just the language of science but also the ideas behind it that can create a barrier between the scientist and the non-scientist. The divergence between scientific language and ordinary thinking was highlighted by Sir Denys Wilkinson, Emeritus Professor of Physics at Sussex University and a former Vice-Chancellor of that university. In a lecture given at Stanford University in 1977 he defined science as being the link between the world of man and the world of nature, but added that the language of man was proving ever more inadequate as a means by which to express the world of nature. He pointed out that many phenomena in physics

. . .are utterly beyond the range of our household experience, on the basis of which we have formed our intuitive pictures of the natural world and on the basis of which we have forged our concepts of how the natural world works, and on the basis of which we have developed our language that we use for transmitting to one another our impressions of the natural world.

Our household language is all that we have to tell each other what is going on out there, whether it is the catching of a bus to which that language is naturally adapted, or whether it is the fissioning of a nucleus or the collapse of a neutron star into a black hole to which it is patently not.

In many ways physics is a special case in that some of its concepts are far removed from what we experience every day, but to some degree all the sciences contain ideas which are not familiar or necessarily reconcilable to common-sense notions of the world.

A third difference, and by far the deepest, is the self-contained nature of science, not in its outcomes but in its processes. What matters in pure science is the truth of a theory, which can only be tested by reference to other theories or by experiment; and in applied science what matters is whether something works, whether it is truly practical. The search for truth lies at the heart of science and, as many scientists, such as Jacob Boronowski, have rightly argued, this dedication to truth is what makes science a civilising influence. It civilises in two senses: it casts out ignorance and superstition, and by its technical and medical fruits it ameliorates and refines our existence. But these external aspects do not affect the essentially self-contained nature of its practice. Whereas the laws of science can only be judged by the criteria of testability and falsification, the rules of the legal or accountancy professions, for example, have to be judged by two criteria – consistency with existing legal or accounting laws and their effect on other people. This second criterion is what makes science so different from most professions. In law, accountancy, business or the media intellectual ideas and pursuits have to be altered to suit the needs of people. In science they do not, nor should they. If we plead that the sun goes round the Earth, as the Catholic Church once insisted, science cannot stand on its head merely to appease us. On the other hand the scientists' disposition to believe that once they have stated a thoroughly tested proposition that is the end of the matter is very far removed from the way of thinking of non-scientists and non-scientific activities.

Another aspect of the same issue is that many scientists are most at home when they are dealing with certainties. This may seem a perverse view of a profession that wrestles with theories of probability and uncertainty, that enjoys exploring the unknown and that owes its extraordinary success to pursuing those activities, but it is the natural consequence of the need for predictively correct answers.

One example will illustrate the point. Professor Tom Blundell, now Chairman of the Agricultural and Food Research Council, once considered going into politics. He was at the time in charge of the planning department of Oxford City Council. He thought a good deal about a political life but decided to stay in academia for the following reasons:

> The problem with politics is that it is very difficult to make decisions on the data available because it is never complete and because there are always pressures on the data that you have. In science there is an emphasis on waiting till one has all the data before one makes a decision. I'm still basically a scientist who likes to say, I need to do just another experiment before I make up my mind.

In professions outside science practitioners have to make decisions with very incomplete information. The chairmen of most large companies say that an essential prerequisite to do their job is the ability to make decisions on the basis of imperfect information. Some people may retort that it shows; but in a very fast-moving world there is rarely enough time to get hold of all the information. So some decisions are inevitably made at least partially on intuition. This is anathema to a scientist and fortunately so. We expect our scientists to be sure of their facts before making statements.

Increasingly, however, scientists are being asked to make pronouncements without access to as many facts as they would like. The usual reason is that the problem is relatively new and therefore any scientific conclusion will lie between the tentative and the probable. An example is the hole in the ozone layer. We may not know for another thirty years all the causes and the consequences of this, but it has become a political issue and therefore one on which the public ought to have an opinion. So one reason why we need to bridge the cultural divide between science and non-science is that so many problems today have a scientific aspect. If we want to manage the resources of the world better, and if we want to understand the natural balance behind those resources, we need to bring science to bear upon the problems. At the same time we need scientists and non-scientists to talk to each other. It is no good just leaving it all to the scientists, we have to be able to understand at least something of what they say. If there was greater communication between scientists and non-scientists, the expectations which we have of science might be more realistic. Many scientists believe that the public is perhaps too overawed by science, that people expect scientists to be able to solve big problems quickly.

Two things need to happen if the gap between science and the arts is to be bridged. One is for scientists themselves to communicate more. As Blakemore says:

> We must be available to the media and not live in our ivory towers. Yet academics have not encouraged each other to be involved in the media. Many people in the media are open to, and genuinely interested in, science, and most will report you accurately.

The second change is for the media to bring science more into the mainstream of their coverage. The tendency to limit the reporting of science to specialised journals or niche sections of newspapers only reinforces the separateness of science. Wolpert asks why scientists and their work are not included in the coverage of general culture. Why, he asks, are scientists rarely invited to general discussion programmes on television and radio? Why indeed.

Most efforts to unite the two cultures are sapped by the pervasive social snobbery that still exists towards science in Britain. This is where Britain is so very different from many other countries. Professor Connerade, whose background is Anglo-French, has always been struck by our slightly superior, condescending attitude towards science. He says:

> This country is immensely practical and always stresses its practicality, but only up to a point. The British do not believe that you should know anything about what it is you're doing. The expert is the last person you should trust to do anything and he shouldn't be allowed to have an opinion because that's above his status. All that is connected with an aristocratic idea and it comes about from having a distinction between rulers and ruled. If it were just a question of being clever and learning about something in order to make decisions and be in a position of influence then, a clever person, from whatever class, could do it, but then the ruling class would be exposed to invasion by all those clever people outside. So the emphasis remains on background rather than on ability; you can be a good scientist but you still might not be of the right class.

This is an opinion shared by Austrian-born Nobel prizewinner Max Perutz:

It's an old British tradition. The people in the humanities have been regarded as carriers of civilisation and the scientists have been regarded as plumbers!

Connerade says that we see the embodiment of this tradition in 'the endless number of ministers who have no knowledge of science', adding:

> It is very noticeable that the first time they speak in public they apologise for not being a scientist, and they do it very publicly. But it's not a real apology: it's an essential qualification for the next job.

> The example set by ministers is followed by others in the government machine. In the civil service there are people who hide the fact that they have a science degree, because they feel that if it was known they wouldn't be allowed anywhere near a department which has to make decisions concerning science.

Connerade contrasts this practice with that of Germany, where the Ministry for Research and Technology is full of people with science degrees, some of whom are ex-students of his colleagues and collaborators in the University of Bonn, where he worked for many years. In governments such as those of Germany and France, scientists are part of the fabric of decision-making – they are not housed in a separate unit like unreliable servants. This means that scientists and technologists in those countries can play a natural part in dealing with the major issues of economic and industrial policy. In this country scientists have played a very small part – at least directly – in government policy-making, and this has been very much a consequence of our two-culture attitudes.

Many scientists, such as Professor Denis Noble, are increasingly saying that this cultural problem impacts on our ability to create wealth. The argument that scientific endeavour influences economic progress is a contentious one, and forms the subject of the next chapter.

3

SCIENCE AND THE NATION'S PROSPERITY

Economists have had much more success in dealing with the consequences of technological change than with its determinants.
Nathan Rosenberg in 'How exogenous is Science?' in his book, *Inside the Black Box: Technology and Economics.*

Rosenberg, who is still studying the relationships between science, technology and economics, recalls the view of another US economist, Stanley Kuznet, who pointed out that 'the distinctive feature of modern industrial societies is their success in applying knowledge to the economic sphere, i.e. knowledge derived from scientific research.' Rosenberg rightly says that this view is 'disconcerting' to economists, for if it is correct, then 'the major determinants of a central economic phenomenon (technical development) lie outside the economists' range of analysis.'

If American economists have not been very adept at understanding what determines technological change and in appreciating the science and technology that lies behind it, is it any wonder that we in Britain have not appreciated it either, for we live in a country where policy-making is dominated by economists who are largely influenced by the writings of their colleagues. There is scarcely a book by a British economist which mentions science in a significant way; this is, of course, just another reflection of our culture's attitude towards science.

The connection between scientific endeavour and economic success is a controversial one. The first problem is how you define science. Is science to be characterised solely by reference to fundamental research, and if so should that include basic research done in industry; or should we widen the definition of science to embrace applied science as well? Ought we perhaps to see science as a circular process whereby developments in industry influence research in universities which, in turn, feeds back into industry? How much 'science' is there in the development of a product after the research has been completed? What do we mean by a scientist? Is an engineer a scientist? These issues of definition and interrelationships are complex, but we tend to over-simplify them by putting science into a compartment separate from technology whereas they in fact have a symbiotic relationship. In a modern chemical or pharmaceutical company, for example, the symbiosis between science and technology, drawing on ideas from within and from outside, is very apparent. The same was true two hundred years ago, when we were leading the industrial revolution.

The second problem is how you define success. When scientists and industrialists argue for more spending on science on economic grounds, they are asked to produce the evidence which shows that this will benefit the country. When George Walden, a junior minister in the Thatcher government, stated in 1986 that the period of highest science spending – the 1960s – coincided with poor economic performance, he was speaking for many people.

A number of scientists and businessmen have questioned the thesis that more science leads to better economic performance. Back in 1975 the government's chief scientist, Sir Ieuan Maddock, cautioned against being dazzled by the attractions of fundamental science. He argued that a major part of our GNP comes from activities that are not particularly sensitive to technological advance and are only marginally affected by research and development, and that we could become preoccupied with the ceaseless creation of new technologies. He concluded by saying that what we needed to do was to put more effort into applied science, and to put more money into better design, manufacturing and marketing and improved management. His words still find an echo among those businessmen who say that what we really need is more innovation, not more science. Even Sir Denis Rooke, ex-Chairman of British Gas, in his opening address to the 1990 conference of the British Association for the Advancement of Science, admitted that 'a huge improvement in industrial innovation could well take place using the fund of scientific knowledge that we already possess.'

The argument made by Walden and others like him is extraordinarlily over-simplistic and, to put it bluntly, naive. One might just as well ask, whether we would have been better off if we had spent less money on science in the 1960s. The answer to that is that we would probably be even worse off. Many of today's leading scientists in industry and the universities, and a good many of the technical advances of the last fifteen years, were products of the science boom of the 1960s. Moreover, as Sir Denis Rooke himself says, while it is true that we need a greater degree of innovation among today's managers, that innovation is usually science-based. A further point that is often missed in this debate is that if you want to innovate you have to have scientifically-trained people capable of intererpreting new ideas, capable of choosing between one possible innovation and another and capable of explaining how that inovation can add value to a company's performance. Skilled scientific manpower can only come from a thriving science base which is keeping abreast of new knowledge.

While the semantics are important, the discussion of definitions can distract attention from some of the salient evidence which does show that the application of science boosts business and economic prosperity. It is at company and industry-sector level in Britain, rather than at a macro-economic level, that one can find evidence which demonstrates the importance of scientific capability as a contributor to industrial success. There is also evidence from overseas. Here are some examples.

In a lecture to the British Association for the Advancement of Science in August 1991, Professor William Kay of the London Business School identified the top UK companies which had created the most 'added value' in the 1980s. 'Added value', he explained, 'measures the firm's contribution to the national and international economy, and the loss which would result if the firm were broken up.' The top ten companies which 'added the greatest total amount of value,' he listed as follows: British Telecom, British Petroleum, BAT Industries, British Gas, ICI, Glaxo, RTZ, GEC, BTR and Grand Metropolitan. There were a variety of reasons why these particular firms were in the list. Many of them featured, as Kay said, because 'they are large rather than because their competitive advantages are strong', or because 'they have great strength in a market niche'. The interesting point is that for a majority of these companies science-based technology is an important factor in their success.

Kay was careful not to draw any simple correlation between science-based innovation and commercial success. Citing EMI's failure to benefit from its lead in medical electronics technology in the 1970s, he pointed out that for a corporation to prosper it had to create 'appropriable added value' – i.e. it must create added value that it can itself exploit and not leave it to be exploited by someone else. Nevertheless the fact remains that these successful 'added value' companies are for the most part built around various science-based technologies.

Science and technology-driven companies also feature prominently among *The Times* top 1,000 industrial companies. Of the top 50 companies in *The Times* list, 58% are technology-based, i.e. technology is an important component of their business. When one looks at Britain's major exporters the significance of science and technology-based companies becomes particularly apparent. In 1990, 76% of Britain's top 50 exporters were science and technology-based, and among the top 10 exporters the proportion was 90%. The success of leading exporters such as ICI, British Aerospace, Shell, Glaxo, Smith Klein Beecham and Courtaulds was strongly dependent on technology and scientific research. In the case of the pharmaceutical and chemical companies, a major part of their profits is directly attributable to a number of key products that began as ideas in research laboratories. Glaxo's Zantac drug is perhaps the best example. At ICI nearly 40% of the profits come from a handful of drugs produced by its pharmaceutical division.

The British industrial sectors that have performed best over the last twenty years have been chemicals, pharmaceuticals and oil. They have generally invested a higher percentage of their profits in research and development and have given scientists and technologists greater status than have their counterparts in other areas of British industry. According to Dr Peter Leonard, an executive at the Chemical Industries Association, the British chemical and pharmaceutical industries spend an average of 7% of their revenue on research and development compared with less than 3% for the rest of manufacturing industry. He adds that these industries have grown faster than their German and American counterparts during the last decade, and that in 1991 they made a positive contribution to the balance of payments of £3bn. This is not to say that spending on research is the only arbiter of success in these businesses, but it is an important factor.

The correlation between spending on research and development and economic success has been shown in a number of reports from PA Consulting Group, Britain's fifth largest management consultancy. In a report in 1988, 'The significance of R & D investment to the UK economy', PA identified the six fastest growing industries as being also the ones with the highest ratio of research and development to output. A report sponsored by PA, British Aerospace and the Institution of Mechanical Engineers in 1989 ('Innovation and investment and the survival of the UK economy') made much the same point, arguing that the UK's poor economic performance between 1969 and 1985 had been partly the result of low investment in research and development. Moreover, compared to its major competitors, the UK was the only country to have shown a decline in the number of research workers in manufacturing.

Since those two reports PA consultants have continued to study the importance of research and development in companies' success. John Fisher, Technical Director at PA's Technical Division, explains some of their findings:

> We did an internal survey recently on the pharmaceutical industry in which we measured companies' research in terms of quantity, quality and timeliness. We found that the companies that had the highest overall ratings came out the best in profitability.

Peter Houzego, a PA consultant specialising in the electronics industry, adds, 'There is a strong economic correlation between science-based product development and performance in the marketplace.'

It is also illuminating to look overseas – at Germany, Japan and the USA. German industry's research and development effort (before unification) represented 1.72% of GDP compared to British industry's 1.14%; research has a very high priority in German industry and the boards of German companies usually have at least two technically qualified directors. British boards are lucky if they have one. Five German companies (Siemens, Daimler Benz, Hoescht, Bayer and Volkswagen) spend £8.5bn a year between them on research and development. This is more than one-and-a-half times what the whole of British industry spends.

Japanese firms have also devoted considerable resources to research and development. We still think of Japan as a technological predator, or 'technological leech', according to The Economist in May 1988, but this attitude really is years out of date. Dr Donald Braben, former head of BP Ventures, rightly says that in many cases it is Western companies who are the leeches on Japanese technology. But, in any case, that criticism, even if once true, obscures the fact that the reason that the Japanese were able to exploit other countries' ideas was that they had a scientifically-literate workforce. Even to be a successful 'leech' you still need a science base.

The reality is that in the last ten years Japanese companies have poured much money into science, and they have put very large sums into funding research in overseas institutions and laboratories. According to the Japanese Ministry of Trade and Industry (MITI), 116 Japanese companies have been financing scientific research overseas. Anyone going round UK universities at the moment will discover that a rising proportion of their funding is coming from Japanese companies. The Japanese are doing this because they believe that it makes commercial sense.

In the USA a study by Edwin Mansfield, an economist at Pennsylvania University, published in 1991, looked at 100 companies in seven industrial sectors. It charted their major technical developments from about 1975 to 1990, and then correlated this data with US government funding of fundamental research. Mansfield found that the rate of return varied between 28% and 40%. This study has much impressed Dr Alan Bromley, White House Advisor on Science and Technology. It has strengthened his advice to the President's office that the USA should invest in areas of strategic scientific importance. One of these is materials science. The US government has pledged $1.6 billion a year for research into this area alone because of the strategic economic and military importance of the field.

There is further evidence from overseas to suggest that our competitors are making research and development a top priority. In a list of companies ranked on the basis of research and development expenditure (compiled by the *Independent* in June 1992) only one of the top forty was British – ICI. Fifteen were American, ten Japanese, and five German. Admittedly one reason is that most of these other companies are bigger than ours, but at the same time it is an indication that companies abroad are putting substantial sums into research and development.

This kind of evidence has been cited, on and off, for more than twenty years. The shelves of the National Economic Development Office, (which functioned from 1962 to 1992) were filled with reports on every conceivable industry all providing the sort of data described above. The same is true of the Cabinet's Advisory Council on Science and Technology (ACOST), whose predecessor, the Advisory Council on Applied Research and Development, also issued studies on sectors of industry which noted declining spending on research. While these reports have gathered dust our industries have continued to lose out in international markets and our unemployment has steadily risen. Some of our high unemployment now is certainly due to our failure to invest in good research in the last ten years and it is foolish to believe that it is solely related to what has happened to the economy in the last two or three years. The impact of research and development spending is felt over a long time-scale, for it often enables a company to maintain or increase market share. If market share is lost through a failure to invest in research then it is frequently very difficult to regain it.

Increasing commitment to research and development among our international competitors is happening at a time when research time-scales are lengthening. Though it is often assumed that industrial research and development times are shortening, this is only true of the later developmental/pre-market stages of the process, where companies are desperate to use management consultancies like PA to help to them to speed things up. But, overall, research times are not shortening at all. As Dr Donald Braben says: 'Everywhere you look, the lead times in any really new area are getting longer. A totally new drug will need fifteen to twenty years to take it to market.' Scientists working in medicine such as Sir John Vane – who appears later – bear that out.

So if this evidence about the importance of research and development is so powerful, why do we in Britain not recognise it? There are many reasons, of which two are of prime importance. One is the fear of failure, while the other is the fear of what the accountants and shareholders will say.

Those who feel that the fear of failure is a reason for doing nothing can cite celebrated instances involving companies that either collapsed or were taken over despite their high investment in research and their achievements at the leading edge of their technologies. In the 1970s it was EMI and Rolls Royce in medical electronics and aero-engines, in the 1980s it was Inmos and Acorn in computers and Plessey and Ferranti in defence equipment, communication systems and silicon chips. These examples are etched into people's memories. What is forgotten is the large number of firms which failed in the 1980s precisely because they had never invested. They were the firms that had been left behind technologically and which were relying more on price than on quality for their market share. When an overvalued pound compounded their difficulties in the early 1980s their technical obsolescence left them with nowhere to go. They collapsed. The exploitation of science-based technologies may be risky but the failure to attempt it is riskier still.

The second fear is rooted in the fact that British firms have given pride of place to accountants. Britain has more accountants per head of the population than any other European country and this partly explains why we rate business performance principally in financial terms. Business success is almost exclusively seen as the efficient use of financial resources. This ignores the role of intellectual capital in contributing to financial success. To understand why accountants are so important in British companies you have to appreciate the financial climate in which they operate. Publicly-quoted British companies are stock-market driven. Nigel Horne, an ex-divisional R & D manager at STC and GEC argues that shareholder pressure is the number one reason why British companies place short-term financial objectives above all others. It explains, he says, why accountants enjoy the pre-eminent status that they have, and why technical goals and technical people are regarded as secondary.

It might be argued that US companies are also stock-market driven, to which Horne responds by saying that US companies are now also affected by short-termism, and that they are worried by the dangers. There is also the argument – often ignored over here – that many of the big US companies enjoy extremely close relationships with US governments, and are able, one way or another, to secure financial support for research, especially if it can be shown to have a defence application. As Horne stresses repeatedly: 'We don't realise that other countries are not playing by our rules.'

British universities seeking joint ventures with British industry see this short-termism all the time. Sir David Smith, a biologist and Vice-Chancellor of Edinburgh University, says:

> We find a terrifying short-termism in British industry. In this university we have a leading centre for speech technology research, which involves linguists and computer scientists. One long-term result of their work could be speech-driven word-processors – you would talk into your word-processor, and it would type – while another is translating telephones, in which you would speak in English, for example, and your opposite number would hear you in another language. British companies have said to us, let us know when you have something that is in sight of development. Japanese companies have made quite a different response: they have given us very large grants. One Japanese company has said to me, we must be where the market is going to be in the next century, so we have to have access to your technologies.
>
> That is not how British companies and their shareholders think. They want at least a 6% or 7% return on their investment and they want to see that within a short time frame.

Smith also points out that this short-termism is changing the way science is perceived by those who fund it. He says:

> If a Minister comes here to see our laboratories, I can't tell him that this year's research will result in value next year, which is what he wants to hear. I'd be very hard pushed to promise that kind of instant result. If I were asked what will be the value of what you are doing in five years' time, then I can give a very different answer.

To really appreciate the role of science in our economy we have to widen our perspective. We need to see how the products of a science base flow into the economy and we should understand that the science base is fostered by a circular process: it stimulates technologies and their users, and it in turn is stimulated by them. Take purely fundamental research in universities and other non-industrial institutions. From this source four kinds of outputs flow: ideas and products, people and skills, instrumentation going into the business sector, and technical ideas and staff for the social/community sector. (A government-funded food laboratory is one example of an institution that makes the last kind of contribution.)

Science permeates the economy through a series of pathways. The fact that some sections of industry do not use those pathways very well does not mean that they do not exist. Some sectors of industry commute back and forth along those pathways all the time.

One important source of new ideas for industry is university research. In chemicals, oil and pharmaceuticals, the universities and industry have forged very close links, which explains why it is in these sectors that the strongest advocates of government support for science are found. ICI's top management have repeatedly drawn the link between high-class scientific research and maintaining a major position in world export markets. But as Sir John Harvey-Jones has pointed out, this research has to be fed with new ideas and new recruits from the universities. If the seedbed of academic research is threatened, then the ability of companies like ICI to compete in world markets is undermined, he says. So in this sense national funding of science does make an impact on the economy. The same message has been expressed to me by managers in other British chemical and pharmaceutical companies.

One area of university science that has a significant potential for the economy is materials science. Colin Humphreys, Professor of Materials Science at Cambridge and Chairman of the Materials Commission at the Science and Engineering Research Council, believes that our capability in materials science will be a deciding factor in our economic performance in the 1990s. He points to the fact that the USA and Japan are pouring huge sums of money into this sector. He says:

> Advances in materials now underlie many product developments and are a strategic area of competition. The next generation of aircraft, motor vehicles, semiconductors and health products are being designed around new materials. They are more sophisticated, often stronger and lighter than anything we are using now. They have become crucially important in aircraft and vehicle design. An aircraft engine that can operate at a higher temperature is more efficient but in order to have that you have to have tougher heat-resistant metals. A lighter airframe will reduce fuel consumption. So we need lighter, tougher metals. The same arguments also apply to motor vehicles.

The science base is also important for the skills that it provides to the rest of the economy. Brian Richards, Chairman of British Biotechnology, says that he and his colleagues deliberately sited the company outside Oxford University 'in order to be near one of the leading centres in molecular biology and biochemisty in Britain'. Research directors in those companies which do rely on scientific ideas for the development of their products and which do invest substantially in research echo the underlying sentiment.

A strong scientific skills base is particularly important in the development of the newer, high-technology companies. In the 1980s many of the new high-tech companies that were rising to prominence were developed by people who had come out of scientific research departments during the 1970s. Sir Peter Michael, founder of Micro Consultants which he merged with United Engineering Industries, says that many of his best scientists came out of advanced research centres. The experience that they had gained in pursuing first-rate research was directly applicable in commercial circumstances. To have been able to tap into that expertise was immensely valuable, he says.

The value of a scientifically skilled workforce becomes apparent if we look at how our industries originally came into being. An interesting view on this comes from Professor Peter Mathias, Master of Downing College, Cambridge. Mathias is both an economic historian – who has written widely on Britain's industrial revolution – and a recent member of the government's Advisory Board for the Research Councils (the main body advising on our nationally funded science research). His views on the factors that helped to make our industrial revolution happen are very pertinent to some of our circumstances today.

Mathias argues that the industrial revolution owed as much to a growing scientific sub-culture and to improvements in education as it did to particular discoveries and inventions. In the eighteenth century the direct linkages between science and innovation were rare, he says. Notable exceptions were the fledgeling chemical industry (dyeing and bleaching) and the developments of steam power. What was very characteristic of this period was a sub-culture of scientifically curious citizens. This sub-culture centred around local societies up and down the country, dedicated to literature and natural philosophy, which brought together amateur scientists and gentleman-manufacturers. Many of these societies had the specific aim of popularising science and using scientific knowledge for practical ends.

> These societies grew up spontaneously and spread to an extraordinary degree. They corresponded with the Royal Society in London and they were part of a culture that encouraged a scientific mode of problem solving. Science became a legitimising activity.

Mathias also points out that the diffusion of new technologies during the early industrial revolution involved a relatively small but crucially important group of highly skilled artisans. The early entrepreneurs were well aware of the importance of their skilled men and did their utmost to prevent them from leaving their employment.

A significant development, from the early 1800s onwards, was the rise of technical colleges which gave education to literate artisans and the rising factory owners. Mathias explains:

> By the mid-nineteenth century every mining centre and centre of engineering in

England and Scotland had given rise to local colleges which provided technical education and which were sponsored by local firms responding to local needs. Virtually every regional centre, including Cornwall, had some college committed to scientific training and to engineering. They grew outside the university system and they have been quite unfairly ignored.

The linkages between scientific knowledge and industrial growth have been underplayed in Britain because people have on the whole ignored many of these provinical colleges such as Firth College in Sheffield. It has only been fairly recently appreciated that there were close links between Firth College and the steel mills of Sheffield. Many of the links were originated by the professor of Metallurgy at Firth College. He acted as technical consultant to local steel firms with whom he had to sign confidentiality agreements, which is why he never published anything and which is why his influence has not been recognised. He is not an isolated example. Today many of these colleges have become science-based universities; Sheffield and Strathclyde are examples.

An important message that comes from economic historians such as Mathias is that economic growth depends upon a skills base as well as a knowledge base. Scientific ideas will not take root unless there is a skills base that can exploit them and unless there is an appropriate level of scientific knowledge among the business community so it that can understand the new ideas.

The third way in which scientific activity can feed into economic performance is through advances in research instrumentation. This particular interaction between science and technology 'receives virtually no attention', says Rosenberg. Addressing the twentieth anniversary conference of the Science Policy Research Unit in 1991, he suggested that the movement of new instrumentation technologies from the universities out into industry had been an output of basic research that had in fact been 'of great significance in specific sectors of the economy'. He went on:

> Much, if not most, of the equipment in an up-to-date electronics manufacturing plant had its origin in the university research laboratory. In this sense, scientific instruments have been transformed into industrial capital goods.

He cited four examples, three of which have had a direct impact on the semiconductor industry: ion implantation, phase-shifted lithography and synchrotron radiation. All three techniques were devised in the laboratories of research physicists and the first two have already been used in the manufacture of integrated circuits, while the third is being vigorously developed for this purpose by Japanese firms and by IBM. The fourth example is the scanning electron microscope, which

> . . .has migrated from its university origins to the world of manufacturing technology. It has become an indispensable measurement tool in microelectronics fabrica-

tion plants, where the elements of memory chips are now at a scale too small to be resolved with optical microscopes.

Rosenberg concluded:

> When the context of discussion is the economic consequences of science, there is no obvious reason for failing to examine the hardware consequences of even the most fundamental scientific research.

It is not just in manufacturing industry that ideas originating in science can find useful applications. Science feeds into the service sector, into the communications and transport infrastructure and into the social sphere. Credit-card banking with its holographic cash cards, electronic point-of-sale check-outs and computerised methods of buying products, computerised reservation of plane tickets and the ubiquitous fax machine are all based on technologies which have derived from science.

In the social sphere, concerns about health, food safety and pollution can only be effectively resolved with scientific knowledge. Scientists may not make the policy, but they are key advisors. When there was alarm about mad cow disease or listeria it was food scientists who were brought in. When the water supply in the polluted Camelford area was cleared as being safe to drink again, it was the scientists who advised the local authorities. Issues of domestic or business safety often involve scientists, too. For example, such questions as when is a piece of clothing flammable or when is a structure unsafe etc. may seem obvious and amenable to fairly simple answers but frequently require a scientific input.

There is yet another way in which scientific ideas can feed the economy, for much scientific knowledge is influential beyond the immediate context in which it was formulated. Keith Pavitt, a researcher at the Science Policy Research Unit, has pointed out that 'science-based technologies diffuse better to other sectors of economic activity,' than do non science-based technologies. Synthetic fibres stimulated changes in textile machinery making and advances in electronics stimulated developments in mechanical and electrical engineering. Thus a development in one area can easily bring about changes in another.

By the same token, developments in industry feed back into fundamental science. The commercial and technical needs of industry put enormous pressure on scientists to come up with solutions, as Rosenberg made plain in his book, *Inside the Black Box*. He returned to this theme in his lecture at the 1991 conference:

> Scientists in industry are inevitably confronted with specific observations or difficulties that are extremely unlikely to present themselves in a university laboratory: premature corrosion of an underwater cable, unidentifiable sources of interference in electromagnetic communications systems, or extreme heat generated on the surface of an aircraft as it attains supersonic speeds. . . . In this respect, the industrial

research laboratory may be said to have powerfully strengthened the feedback loop running from the world of economic activity back to the scientific community.

Scientific knowledge permeates our economy, circulating through the different business sectors and at the same time moving back and forth between research and industrial centres. The quantity and quality of that scientific knowledge is important. This means that the diffusion and application of that knowledge has to be improved. However, this will not be possible without a strong science base.

The importance, from an economic point of view, of maintaining a dynamic science base is strongly made by Derek Roberts, Provost of University College, London, and former research director at GEC and Plessey. He believes that it is extremely misguided for politicians to undervalue the benefit of science to our economy. He says:

> Spending money on science doesn't guarantee economic growth because of everything else you need to do to achieve business success, but equally if you don't put money into science you certainly won't get any economic growth.

But it can still be asked whether, if we are to be economically prosperous, we must do the leading-edge research ourselves? There is a view popular among politicians and civil servants that we could buy in research from other countries, letting them take the financial risk. Roberts thinks that proposition wildly unrealistic for three reasons:

> Firstly, if we let other countries do the basic research where are UK companies going to recruit people with a good understanding of advanced research? Secondly, if we put that degree of distance between ourselves and our competitors and create barriers between the exploitation of research and where it's done, it is difficult to see how we are going to get into positions of market leadership. We'll always just be following behind. Thirdly, if this is going to be the UK strategy and we leave research to Germany and Japan, you can judge for yourself which country will be more successful. It won't be us. The proof is there already. Just look at the German and Japanese record of research investment over the last twenty years. To ignore all that just isn't rational.

The strategic importance of science to the economy has also been stressed by Sir John Cadogan, BP's Research Director. Addressing the Royal Society some years ago, he said:

> It is worrying that this highly educated nation shows signs of forgetting the lessons of history, which are that scientific discoveries are at the roots of all economic progress, and that they will continue to be made. Curiosity-driven research in strategic areas such as catalysis, materials, solid state physics, optoelectronics, biomolecular science, synthesis, will pay off, and the pay-off will be substantial even if it is measured only in terms of the production of qualified people.

These views lend weight to Rosenberg's conclusion that science and technology policy should be seen as 'an aspect of economic policy-making'. This aspect has two parts. One is the creation of an environment in which the science base can flourish, and the second is 'providing a structure of economic incentives and rewards that are supportive of the rapid diffusion of new technologies, once they have been developed'.

In the last ten years, we in Britain have, if anything, moved backwards on both counts. The recent attempts we have made to stimulate technology transfer or encourage market driven science have either been negated by other economic decisions or they have been totally misconceived.

Part II
LIVES IN SCIENCE

4

MEDICAL SCIENTISTS: THE MEANS TO HEAL

There is no such thing as the scientific method. . . . A scientist who wants to do something original and important must experience, as I did, some kind of shock that forces upon his attention the kind of problem that it should be his duty and his pleasure to investigate.
Sir Peter Medawar, immunologist and Nobel prizewinner, in his autobiography, *Memoirs of a Thinking Radish.*

Sir Peter Medawar became a leading authority on skin grafts and his work also laid the basis for organ transplants. His interest in how the body 'discriminates between its own and other living cells', came about largely by accident. In about 1942, as a very young scientist, he saw a German airman crash about two hundred yards from his home. Within a few hours he was called in to advise on the treatment of the third-degree burns which covered 60% of the man's body. This is the 'shock' to which Medawar is referring to above. He had already done some research on war wounds and on burns in Professor Howard Florey's laboratory, but the airman's burns were so bad that they required skin grafts. At the time there was little knowledge or understanding of how skin grafts worked. Medawar now set himself the task of studying skin grafts which meant investigating how the body's immune system worked.

Medawar's work and that of many other physiologists during World War II was supported by government funding. This money was extremely well-spent since it not only saved lives but it also advanced many fields of medicine. After the war Medawar, and most of his colleagues in medical science, continued to be financed by public funds.

British governments have been funding medical science since 1913, when the Medical Research Committee, now the Medical Research Council (MRC), was founded. It now is the principal government agency funding medical research. Since that time MRC-funded research has led to a host of discoveries, including the cause of rickets, the isolation of the influenza virus, various antibiotics, the link between smoking and lung cancer and early influences on heart disease. The

MRC has also aided the development of research into magnetic resonance whole-body scanning and has been a major funder of molecular biology. In addition, it has funded research into veterinary medicine which among other things has resulted in a number of vaccination and immunisation procedures. Today, the MRC spends just over £200m a year on research covering medicine, molecular biology, genetics, mental health and nutrition. Nearly 60% of this money goes directly to 56 MRC-run establishments while the rest goes to individual researchers in universities and hospitals.

After 1945 MRC funding for university and hospital medical research grew rapidly, partly as a result of the creation of the National Health Service. At the beginning of the 1980s, the MRC was still the major provider of funds for medical research with charitable foundations and pharmaceutical companies contributing the rest. However, by the end of the 1980s charities and industry were funding about 55% of Britain's medical research. The Wellcome Foundation alone now funds about 20% of the £500m a year that is spent on the field. For the very obvious reason that health is a prime concern of society, medical research is faring better than some other parts of science, yet even in this area scientists are experiencing difficulties.

The five scientists in this chapter are all working in important and interesting fields of medical research. They have been stimulated by the kind of practical problems that inspired Medawar and also by a recognition that current theories in their field were either misconceived or incomplete. All of them have been engaged in basic research, asking fundamental questions about disease and body chemistry. A number of drugs have emerged and will emerge from their current work. Their interests vary but overlap, with cardiology being a common theme. Sir David Weatherall's interests include genetics as well as medicine. Other preoccupations of these scientists are neurophysiological disorders such as Alzheimer's and Parkinson's diseases as well as afflictions such as asthma and schizophrenia. The scientists' funding comes from a variety of sources. Each in his or her own way is regarded as an important contributor to their field.

The first two scientists, Professor Denis Noble and Sir John Vane, are both engaged in leading edge-science. Each expects that their work and that of their teams will eventually result in new treatments. Indeed, in the case of Vane this has already happened; one of his discoveries led to a new hypertension drug and to another new product, due out in a few years, which should help to inhibit thrombosis and hardening of the arteries.

Denis Noble is Burdon Sanderson Professor of Cardiovascular Physiology at Oxford. He studied to be a doctor at University College Hospital in London and then switched to medical science, largely as a result of encouragement from Dr

Leonard Bayliss, son of the famous Sir William Bayliss, a leading physiologist in the early part of this century. He was, he says, seduced from clinical medicine to science by the sheer enthusiasm of Leonard Bayliss, but in recent years his interests have moved back towards medicine as some of his work has pointed the way towards tangible applications. Noble's lifelong interest has been the study of the heart. Like all scientists, he has drawn on the work of others and his research has been complemented by advances in other sciences, notably mathematics and computing. Most people recognise that mathematics is important in physics and astronomy but it also has an increasingly significant role to play in subjects such as biology and medicine.

Noble was introduced to cardiac physiology at first hand by Professor Otto Hutter at University College, London, who supervised his postdoctoral research. Hutter's interest was heart tissue, in particular studying the regular excitation in the muscle fibres. The heart muscle is quite unlike any other muscle in that it is not dependent on an outside stimulus from the nervous system to make it work. A heart will beat after it has been removed from a body and if the pacemaker is cut up into pieces those pieces will also beat. As Noble explains: 'We know now that even the individual cells in the heart beat. There is a rhythm generator in each cell like a sort of clock.' The question that puzzled Noble was: 'What makes the clock tick?'

At the time that Noble was working on this question (1959/1960), Sir Alan Hodgkin and Sir Andrew Huxley – who won a Nobel prize together with Sir John Eccles – had already built a theoretical model of how the nerve cell, or neuron, works. This model, usually called the Hodgkin/Huxley model, showed that ions (electrically charged atoms) of certain chemicals are exchanged across the nerve cell's membrane and that this excites or inhibits the cell's activity. In thinking about the heart, Noble took the Hodgkin/Huxley model of the neuron which he modified, and then using the experimental information derived from his work with Hutter built up a model of the rhythm of the heart.

The model was a mathematical one and was so complex that all the calculations had to be fed into a computer. Michael Bernal of the London Computer Centre, the son of the well-known scientist J. D. Bernal, thought that trying to make a computer-based model of the heart with the equations that Noble devised was not at all likely to work, and he was reluctant to allow the young Noble valuable computer time during the day. Noble persisted and was given two hours on the University computer between 3 o'clock and 5 o'clock in the morning.

With the help of the computer Noble succeeded in producing the first mathematical analysis of the rhythm of the heart. He was thrilled. 'It was like finding gold, what any young researcher dreams of.' He had opened up a whole field – computer modelling of the heartbeat – which he has been developing ever since. There were important implications behind the mathematical model. Noble says:

> I was able to show for the first time that all the individual bits of the heart mechanism
> that were there in the heart cells are totally adequate to make it oscillate.

The model partly rested on studies that he and Hutter had done on the role of
chloride, sodium and potassium ions in the cardiac muscle. Noble focused on the
potassium ions. The exchange of electrically charged sodium and potassium ions
passing in and out of the cells through the surrounding membranes is what causes
the heart to beat. Noble's contribution to this understanding was to show how the
potassium moves across a heart cell's membrane through what are known as
potassium channels. Up till that time the way that sodium worked was pretty well
known but 'potassium had been a big hole in our knowledge'.

As Noble acknowledges, he was unusual in that he made an important
discovery early on in his research career.

> Research is by and large a grind. There is a lot of disappointment, ninety-nine times
> out of a hundred you're wrong. What you do is often very critically reviewed and
> people will say that you need to do extra work to get your theories right. So a lot of
> research is drudgery. But what keeps us all going is the extraordinary thrill when you
> get on to something. You can't contain yourself. I can still remember the excitement
> of discovering pacemaker rhythm in the model, particularly since Michael Bernal
> had been so sceptical.

Some of this work was to lead to a new family of drugs. One of the biggest-selling
drugs in use today is a potassium channel blocker called amiodorone which is used
to control an irregular heartbeat.

A further discovery in which Noble played a key part was the discovery of the
role of what is known as the sodium-calcium exchange in the heart's electrical
mechanism. As a concept it is similar to the sodium-potassium exchange in which
sodium moves across a heart cell membrane in one direction and potassium moves
across from the other side. Sodium and calcium move in and out of heart cells in a
similar fashion. The actual discovery of the sodium-calcium exchange came from
a German scientist, Reuter. Like most other scientists, Reuter believed that the
exchange merely kept the two chemicals in balance in the heart and that it had no
direct electrical effects. But Noble, who was still working on large mathematical
models of the heart, disagreed. He says:

> I was convinced that we had seen the electrical effect. We observed the ripples of
> electric current during the heart beat which is what you would expect if there was an
> electrical effect.

Noble's model required that more sodium than calcium pass through the sodium-
calcium exchange. However, it was a long struggle to get this thesis accepted. For
about five years, Noble and his co-workers (particularly Dr Don Hilgemann in the
USA) championed their view against the received wisdom of the time. They
eventually succeeded in overturning it when other scientists (including one of his
own students), using even more direct methods, demonstrated the phenomenon
unambiguously.

The sodium-calcium exchange is now seen as the basis of a common and often benign arhythmia in the heart. When your heart misses a beat this is usually caused by the sodium-calcium mechanism working just that little bit harder. This can be induced by a number of causes, of which the most frequent is a temporary lack of blood.

Today one of Noble's principal interests is what causes fatal heart attacks, not in the sense of what triggers an attack, but what it is that induces such electrical chaos that the heart's whole mechanism is destroyed. In a minor heart attack some of the heart tissue is damaged but the heart recovers. In a fatal heart attack the spasm is so chaotic that all the cells are killed. So the core question, says Noble, is 'what lies behind the asynchronomous event that kills?'

To try and answer this question, Noble and a team of collaborators in Oxford and in the USA (where he is collaborating with Dr Rai Winslow) are seeking to construct a computer model of the entire heart. A working model already exists which is added to year by year. It is a massive task – the heart has about two to three hundred million cells – and Noble believes that the project will last at least ten years. He says:

> We may not solve the problem in ten years, but we may lay the foundations for others. It's fishing in the dark to an extent because this is not going to be a case where a single answer is found. My hunch is that we are dealing with an integrative property of a very complex structure, and it will be by understanding the mathematics of how that works that we will get the clues as to how fibrillation gets triggered.

The research demands massive computer power and data is regularly fed into a 'parallel' computer sited at the University of Minnesota. A parallel computer is a machine that includes a number of processors, the bits that actually do the number-crunching, running in parallel, rather than a single processor as in most computers. Parallel computers are particularly suited to simulating the behaviour of large numbers of interacting units, like the cells in the heart.

Noble would have preferred to have had access to a British computer. But none has been available. Oxford University has been waiting for years for financial backing to obtain its own parallel computer, which it has now at long last received with the setting up of Oxford Parallel. But that, for him, has come three years too late. It is also a disgrace as he explains:

> This university has one of the best teams in the world in computing and it has won a Queen's Award for Industry because of its contribution to the development of Inmos and the development of the transputer.

(Inmos was founded in the 1970s as a mass-producer of silicon chips and it developed the advanced 'transputer' which can provide the basis for parallel processing. Funded by the National Enterprise Board, it was sold off by the Conservative government in the 1980s to Thorn-EMI, which later sold it to the French electronics giant Thompson.)

The lack of a parallel computer in Oxford has been damaging, says Noble.

None of my group is getting the experience to work with parallel computers, only me. I communicate with Minnesota University from my office by electronic mail. So this means that I can't train people for the pharmaceutical industry with the knowledge of parallel computing where they could use our research results on cardiac arhythmias to design new drugs.

Noble adds that this work is not just important from a scientific or even a medical point of view but also from an economic viewpoint. Waiting in the wings are the drug companies who will join forces with those researchers who are reckoned to be closest to finally cracking the secret of heart arhythmias. If US researchers gain a significant edge then it is likely that US drugs companies will come in to exploit the ideas, which would be a loss to the British pharmaceutical industry.

Sir John Vane, now Director and Chairman of the William Harvey Research Institute, has been interested in complementary aspects of cardiology as well as other branches of medicine. Vane has been involved in a series of exciting projects which highlight the fact that scientific discovery involves luck as well as hard graft; but, as he adds, luck is of no use unless you have the knowledge with which to exploit it. He won the Nobel prize for medicine in 1982 – with Bengt Samuelsson and Sune Bergsrom – for his research on prostaglandins and for his discovery of how aspirin works. Like Noble, much of his work has been undertaken with colleagues from other countries, and the international aspect of science is highly appealing to him. Unlike Noble, all his current research is funded either by charities or by pharmaceutical companies such as Glaxo, Ono and Parke Davies. From 1973 to 1985 Vane was Research and Development Director for the Wellcome Foundation Ltd, now Wellcome plc.

Vane's first discovery shed light on high blood pressure and resulted in a drug that counteracts it. He was at the Royal College of Surgeons at the time, where he was investigating the renin angiotensin system. This system originates in the kidneys which release an enzyme called renin which acts on a particular blood protein to form a peptide called angiotensin One. (A peptide is a fragment of protein which acts as a chemical messenger.) Angiotensin One is quite harmless, but when it is activated by another enzyme it becomes angiotensin Two, which causes high blood pressure. A critical question at the time (1965) was whether angiotensin caused hypertension physiologically or pathopsychologically. No one knew, says Vane, and it intrigued him.

While Vane was studying angiotensins, a young Brazilian, Sergio Ferreira, came to work with him as a postdoctoral student and brought with him some snake venom prepared from a Brazilian snake, *Bothrops jararaca*. This snake venom was the one from which Ferreira's teacher, Roche e Silva, had isolated bradykinin which is a potent vasoactive peptide. (A vasoactive peptide acts on the blood vessels.) Bradykinin dilates the vascular system. Vane had never forgotten something instilled into him by his first mentor, the famous Oxford pharmaco-logist, Harold Burn, which was that you should never ignore the unusual. So Vane suggested that he and Ferreira look at the effects of the snake venom on the enzymes contained in the renin angiotensin system. A colleague of Vane's, Dr Mick Bahkle, found that the snake venom inhibited the important enzyme that converts angiotensin One into angiotensin Two (which is the one that causes hypertension). This discovery suggested that some of the constituents of the venom – if they could be isolated – could be of pharmaceutical benefit in countering hypertension.

Vane was a consultant at the time to the US drug firm Squibb. He persuaded the company to do some sophisticated tests on the venom in the belief that it would eventually result in an anti-hypertension drug. Squibb's scientists were keen on the idea but the marketing managers 'thought the proposal was crazy', says Vane, because they assumed that any anti-hypertension peptide in the venom would have to be injected by customers – even if it was proved to work. Few people would want to buy something that you had to inject! Luckily, Vane had Squibb's president on his side; the research went ahead. Squibb's scientists managed to isolate an important peptide in the venom, synthesised it and then made a kilogram at a cost of $50,000, and it was proved to be anti-hypertensive. But the marketing people were also proved correct – the drug would have to be injected. Nevertheless, the concept had been proven: the renin angiotensin system was important in causing high blood pressure. The company's scientists were then able to produce a drug that could be taken orally, known as Captopril. Another US company, Merck, went on to develop a similar drug. Vane reflects: 'None of this would have happened if Sergio Ferreira had not come to work with me and if we had not been able to work on his snake venom in my laboratory.'

Meanwhile, Vane was working in another organ – the lungs – which also had cardiovascular implications. He explains:

> There was a new discovery at the time, which we worked on, which was that the lungs were not just a vehicle for taking in air but that they also activated certain hormones and inactivated others. This metabolic function turned out to be con-nected with the endothelial cells. The endothelial cells line every blood vessel in the circulation system.

In particular, Vane was interested in discovering what happens when the lungs go into a state of shock, for example in asthmatic attacks. As he says:

> We wanted to know what it is that causes the shutting down of the circulatory system and the air ways. What were the chemical substances that were released during an attack that caused the asthma?

There were several. They included histamine and two new ones – prostaglandin and a substance that Vane called RCS. Prostaglandin had been discovered in 1935 by Ulf von Euler and was now being isolated by Sune Bergsrom in Sweden, but the conventional wisdom was that it was manufactured in the prostate gland.

Prostaglandin was known to cause muscle relaxation in some instances but to cause pain in others. Its discovery in the lungs was new. We now know that there are several prostaglandins in the body and that one or more of them can be made by most cells.

Again Vane's researches were fortuitously accelerated by another fateful encounter. He was joined by a new student, Priscilla Piper (today Professor of Pharmacology at the Royal College of Surgeons), who was interested in aspirin. Aspirin had been around in tablet form for seventy years and before that its natural/herbal constituents had been used for several thousand years. Yet no one knew how aspirin worked. Piper's interest in aspirin was therefore understandable but the connection between her interests and Vane's seemed remote in the extreme, yet Vane accepted her. He explains:

> My philosophy is to allow people to work on things that interest them, whether or not I think it a good idea. Nine times out of ten I'm right, but they will still learn something and then move on. However, in one out of ten cases, I'm proved wrong which is every reason for giving people their heads.

Piper worked as part of Vane's team and experimented with aspirin, injecting it into lung tissue. She discovered that it prevented the release of the substance RCS. Over the weekend, when Vane was writing up his team's work, it occurred to him that RCS might in fact be another prostaglandin. Moreover, he had recently read a paper by two scientists, Samuellson and Anegard, who had shown that the lungs could contained an enzyme that inactivated prostaglandins. It occurred to Vane that aspirin worked, not as this enzyme did, by destroying prostaglandins, but by preventing their formation. Recent obervations in the burgeoning field of prostaglandins showed that some caused pain and inflammation. Returning to his laboratory on the Monday morning, Vane told his team that he felt he knew how aspirin worked. He took some lung tissue, measured the amount of prostaglandins formed and then tested the effects of morphine and aspirin on the tissue extract. The aspirin prevented the formation of prostaglandins. Further tests showed that it worked because 'it interfered biochemically with the enzymes which synthesised the prostaglandins'. Vane adds:'It meant that prostaglandin was involved in causing the inflammation and pain that aspirin prevented.' There were really two discoveries: Vane and his team had learnt both about aspirin and prostaglandins.

A totally unexpected consequence of this research has been the discovery that aspirin reduces the risk of heart attack, thrombosis and arteriosclerosis – hardening of the arteries. Aspirin does not remove the platelets from the circulatory system, but what it does do is to prevent the platelets – when they stick together – from making a noxious prostaglandin-like molecule called thromboxane which leads to heart attacks and strokes.

However, Vane and his research team – in another major discovery – did find another substance which prevents the formation of platelets. He called this substance prostacyclin. The experiments surrounding it proved very exciting, as he explains:

> We found that if we ground up bits of blood vessels and gave them the precursor of prostaglandins, they would make something that nobody had recognised before and that something – which we called prostacyclin – would cause the arteries to dilate and more importantly prevent blood cells (platelets) sticking to the arteries.

But where did the substance come from? To his surprise Vane found that it came not from the arterial wall but from the endothelial cells that line the arterial wall.

Vane expects that this discovery of prostacyclin will result in an anti-thrombosis drug within a few years. This prompts a reflection on the long time-lag between scientific discovery and pharmaeutical application. He says:

> The timescale is about fifteen to twenty years. We discovered prostacyclin in 1976 but an orally active drug won't be marketed until about 1996. We discovered the snake venom peptide in about 1966 but drugs like Captopril weren't released until the 1980s.

Vane appears to be relieved that he is not dependent on government support for his research. Backed by charitable foundations and by pharmaceutical companies, who are happy to take a long-term view of research, he knows he is able to conduct fundamental research on a five-year time-scale at least and not to have to worry about where the next research grant is coming from, at any rate in the short term.

While Sir John Vane has spent a lifetime looking at the chemical substances in the lungs and the heart, Leslie Iversen's career has been devoted to the chemistry of nerves and the brain. Ever since, as a young man, he read Aldous Huxley's *The Doors of Perception* which recounted the author's use of mescalin, Iversen has been interested in the effect that chemicals have on the brain. Until 1983 he was a full-time academic researcher. Today he is Research and Development Director in the UK for the US drug company, Merck, Sharp & Dohme Ltd.

As with Vane, his work partly rests on the researches of others decades ago. One of these was Sir Henry Dale, the pharmacologist who 'showed in the 1920s that nerves don't transmit to tissues by electrical impulses but by releasing chemicals'. When Iversen was a student it was common knowledge that there were chemical transmitters in the body but it was not proven that chemicals also played a role in the brain. 'We now know of at least forty different chemicals released in the brain,' he says. These chemicals are what are known as neuro-transmitters; they are released at each of the billions of junctions between nerve cells in the brain and act as 'on' and 'off' signals. Scientists' knowledge about neuro-transmitters has grown enormously in the last fifteen years, but even so they recognise that they have yet a great deal yet to discover about this central aspect of the brain.

Iversen's early research involved looking at the process by which a neurotransmitter called noradrenalin was inactivated. The process had been discovered by an American scientist, Julius Axelrod, who was later to receive the Nobel prize. On Axelrod's team was a British scientist, Gordon Whitby, and it was Whitby who, on his return to Britain, first invited Iversen to do research on noradrenalin. Noradrenalin is very like adrenalin in its chemical constitution and also in its stimulatory effect. The research into noradrenalin was to lead to an understanding of how some of the new anti-depressant drugs on the market really worked, and it also fuelled a further generation of anti-depressants.

The research on noradrenalin began with a number of questions. As Iversen explains: 'We wanted to know what happens to the noradrenalin after it has been released, to discover what de-activated it and we followed it by tagging it radio-actively.' He adds:

> There was also a practical aspect: to find out what drugs might influence the mechanism of noradrenalin. If we could identify the particular drugs that have a particular active ability to block part of the noradrenalin process, it might explain how those drugs worked. It proved be so.

Axelrod and Whitby had had the novel idea that after the release of noradrenalin some of it was recaptured, i.e. it was just put into store, as it were, until it was used again. Iversen confirmed this, but he also discovered something else as he explains:

> I was among the first to show that the drugs used to treat depression – amitryptiline and imipramine – all shared the common property of being good at blocking the recapture of noradrenalin. They were enhancing a process that was defective in the depressed brain.

Iversen followed this discovery with a visit to the USA where he worked for a year in the Axelrod group. The following year he worked at the Harvard Medical School, where he researched into two amino acid transmitters (chemical messengers) – glutamic acid, which excites the nerves, and GABA (gamma-amino-butyric acid), which inhibits them. The research – mainly done on lobster tissue – was tedious and extremely difficult. Experiments had to be repeated over and over again 'because the chemicals released were in such minute quantities.' The principal result was to prove that GABA was an important messenger substance inhibiting muscle movement. That research helped to confirm the status of GABA as a neurotransmitter chemical, and has been followed by much subsequent work on GABA in the brains of higher animals.

This kind of research appeals, says Iversen, to people who are easily bored, who are restless, and who dislike routine. But is not repeated experimentation on lobster tissue intensely routine? Not so, says Iversen.

> In basic research, no experiment is ever the same. There is also the excitement that you never know what you will be doing in six months time. The field will move on and you will move with it and you will look back and and say, how ignorant I was six months ago.

When Iversen returned to Cambridge in 1970 he started a group to study the biochemistry of the brain, with particular reference to schizophrenia. The results were not as spectacular as he had hoped, though his team did make one major advance. The research confirmed the hypothesis that an excess of dopamine in the brain is an important factor in schizophrenia. 'We now know,' he says, 'that all the drugs used to treat schizophrenia work by blocking dopamine.' It is also known that an absence of dopamine in the brain leads to Parkinson's disease.

Confirming the dopamine hypothesis – mainly by experimenting on dead brain tissue – was exciting but, as Iversen comments, 'we weren't able to discover the cause of the dopamine imbalance.' Finding causes rather than symptoms is always harder in medical science. This was to prove true again in Iversen's work on Alzheimer's disease. Post-mortem research on the brains of people who had suffered from this disease indicated that they had been severely deficient in another chemical messenger – acetylcholine. But what precipitates the debilitating chemical abnormality?

Iversen, like Sir John Vane, has spent many years studying a group of chemical messengers called peptides. Twenty years ago, it was thought that there existed only a handful of peptides, but as Iversen explains, 'we now know that there is a huge number'. In particular Iversen has shown that peptides affect the brain cells and influence their activity.

At the moment, what Iversen and many others are able to discern is a great variety of information made up of chemical indicators which individually give clues to what happens in the brain. However, they provide only tantalising glimpses. We do not yet know enough about the messengers, or the receptors through which the chemical information passes into the cell. Moreover, scientists are continually discovering new neuro-transmitters and they have come to realise that these chemicals – though they may appear to be similar to others – usually work in very specific ways. The more we learn about the human organism, especially the brain, the more we realise that it is a very precise set of mechanisms.

Like all scientists, Iversen is moved by intuition as much as by the known facts:

> I have a strong feeling that much of mental illness is possibly explicable in very simple terms, and that if only we knew how to unlock the biology of the brain then we could make some dramatic advances in treatment.

It is the relatively new field of genetics combined with molecular biology and chemistry that is helping to drive the next generation of pharmaceutical products. Iversen explains:

> The molecular biologists can come up with enormous surprises for the pharmacologists. They may say to us, you're interested in such and such chemical receptor; we have just discovered the genetic basis for this receptor. More than that, we have found that there are not one but six genes involved. They are all related but they are not the same, because each one specifies a sub-variety of the receptor you are interested in.

The implications of having such specific knowlege is extremely important for a pharmaceutical company as Iversen points out:

> If a molecular biologist comes to us with that kind of detailed information, we can consider designing a drug that only targets a specific sub-variety of the chemical receptor and nothing else.

That refinement means that drugs should eventually be capable of development with few, if any, side-effects.

Iversen stresses the importance of the science base to medicine and to the pharmaceutical industry associated with it. He says:

> For those of us in the drug industry we need a strong science base not only to recruit from but also because it represents a cadre of people with whom we can interact. The people in my research teams are top scientists, and they want the stimulus of being part of a wider scientific community in the UK. This would be true of scientists in any country.

Iversen says he is very concerned about the shrinkage of the science base which he has perceived over the past decade, adding, 'things are not getting better'. He goes on:

> We seem to have a head-in-the-sand attitude. If you don't have top-class well-equipped British research centres, you won't have top-class academics in this country – they will go abroad. If we don't have a reasonably flourishing science base, then we will not get the quality of science graduates and postgraduates that we need.

The success of the British pharmaceutical industry in the last decade has been extraordinarily impressive, but it has not been recognised and, as Iversen says, will not continue without government recognising its responsibilities.

> Of the top twenty drugs on the world market in the last few years, seven were invented in Britain; of the top five, three came from Britain. This creativity is a wonderful asset, but it needs to be supported in terms of research facilities and training. No one in government seems to care.

One development that particularly irks him has been the pressure on firms funding university medical research to pay an increasing proportion of the universities' overheads. It is one thing to pay for researchers' salaries or to pay for equipment, but it is quite another to be asked to cough up and contribute to general overheads.

> The pharmaceutical industry doesn't expect to have to pay for the whole university system. It's not our job to pay for the infrastructure. That's not how other countries do it. The British pharmaceutical company, Glaxo, had the option of building its new research centre in the USA or the UK. It chose the USA. Don't be surprised if other companies follow that example. The cost squeeze in universities has just gone too far, and there's no optimism among younger people in entering research careers.

A similarly gloomy view of British science comes from Dr Annette Dolphin,

Professor of Pharmacology at the Royal Free Hospital, London. She also points to
the low morale among young medical scientists, concerned about how long their
research will be funded and about poor career prospects and to the general lack
of support that many in her field experience. Like Leslie Iversen, she has
studied how chemicals behave in the brain, how neuro-transmitters work
and how a deficit of dopamine triggers Parkinson's disease. Again like Iversen, she spent part of her early career in the USA at the Yale University
Medical School. Her recent work has been researching so-called ion channels.
These are channels in a cell's membrane that conduct ions into and out of
the cell body.

Dolphin's own research spans that of Iversen and Noble in terms of the science
that it draws upon. For Dolphin is both an electrophysiologist (looking at
electrical activity of cells) and a neurochemist (studying the chemicals in the
brain). Her special interest is in studying the pathways through which the
chemicals act. She says she is fascinated by the use of electrophysiological
techniques which enable her to see on a computer screen biochemical events
as they occur.

An interest she shares with Iversen is the study of how the chemical receptors
which lie on the surface of cells behave. In her case she has been looking at how
receptors influence what happens inside the cell. One mechanism is that 'they
couple with calcium channels and modify the ability of calcium to enter the cell'.
Dolphin's studies have unravelled how some of this activity happens. Essentially
she has shown that there is an important intermediary, or what she calls a switch,
between the receptor and the calcium channel. This switch, called G-protein,
shuttles between other molecules. Dolphin's line of study was partly influenced
by two pieces of research on dopamine, one by Iversen and another by an
American scientist, Gilman. The American research was particularly important
since it showed that dopamine worked, not directly on enzymes in the brain, but
through a G-protein molecule. This discovery triggered a question in Dolphin's
mind: 'If G-proteins work in that example might it also be the case for other areas
such as calcium channels.' She showed that a G-protein does connect up with
calcium channels.

Unlike many of her colleagues in medical science, Dolphin is not looking for
clinical or pharmaceutical applications from her work. Applications may well
come, she says, but she does not believe that possible applications should drive
research. 'If you make advances in basic research then that can be used by more
applied scientists.' Some scientists, for example, having seen the importance of
mutated G-proteins in tumours are looking at the possibility of targetting them
with novel anti-cancer drugs. Dolphin has no quarrel with those who look for such
applications, but she emphasises that there is an important place for untargeted
research. It is not a fashionable view. It is certainly easier to get funding for a

project that has clearly identifiable applications than one that has not.

The ability to do serious fundamental research is, she believes, being undermined by the squeeze on funding imposed on the MRC.

> If it wasn't for the Wellcome Trust medical research would be decimated in Britain. The MRC is not supporting medical scientists sufficiently at all. The funding is just not keeping up.

This means that the rate of grant acceptance is lower than it used to be so that 'it is almost not worth putting in an application because of the amount of work that you have to do.' To draw up a proposal for research will involve up to a month's full-time work, she says, adding that, 'It's almost a full-time job getting funds for the people who work with me.' She says that she can see the effect of this uncertainty on her young colleagues.

> I see them not knowing whether to carry on in research or not. The money comes in in small tranches and you never know from one year to the next what the government's science policy will be and therefore how much money the MRC will have. I don't think they feel as excited or as keen as I did when I started out because they don't know if they are going to be able to have a career in medical research. Science now is just not considered important. Unlike other countries where I have been – the USA, France and Germany for example – the government doesn't seem to understand the value of science. It doesn't understand that we're training people for industry for the future as well as developing the subject.

The arguments about the need to fund basic medical research are strongly underscored by Sir David Weatherall, Regius Professor of Medicine at Oxford and Director of the Institute of Molecular Medicine. He believes that the achievements of medical science have been taken for granted and also that politicians have the mistaken idea that the best results in research only come from targeted studies. He stresses the importance of doing research for its own sake, citing the discoveries that led to penicillin to clinch his point.

Politicians invariably want instant results, or at least results in the near future, but when this demand takes the form of setting narrow goals for scientific research the results, says Weatherall, can be very disappointing. He cites Richard Nixon, who after seeing Americans land on the moon, decided that he would put an enormous amount of money into finding a cure for sickle cell anaemia. However, 'the timing for that objective wasn't right', says Weatherall. He explains:

> We were so far from having any real understanding of the basic mechanisms of sickling, and why it varied so much from person to person, that nobody had the appropriate questions to ask. Therefore the money was wasted. There is a time in science when everything has come together and it is appropriate to make a major push into a field, e.g. the human genome project at the moment; the technology and the science are right and it is clear that the thing is do-able. So you can have targeted

research, but it's got to be realistically achievable, given what you know and given the technology that you have at the time. But in the main, science hasn't progressed by setting specific targets. To think in those terms is to be completely ignorant of the way science works.

He continues:

If you look back at the principal advances in modern medicine, they have come from basic work much of which didn't look as if it was going anywhere at the time. Then suddenly a number of things come together and you end up with a major advance. Penicillin was based on a hundred years of work – the ideas behind it didn't start in the 1920s which is what is popularly thought – and many of the constituent bits of research that led up to penicillin were undertaken just for their own sake, not to produce an antibiotic. For example, the Oxford researchers whose work contributed to penicillin, were interested in cell walls and how some substances could stop bacteria from dividing. When they started this work they had no idea that their research would lead to a major antibiotic.

Weatherall adds that it was only as they went along that they started to look for promising antibacterial agents.

Weatherall is a haematologist (an expert on the properties of the blood and blood-related diseases) and a geneticist. He is principally known for his work on the basis of a particular blood disease – thalassaemia – which was known to have genetic roots. The cause of thalassaemia is defective production of red blood cells (haemoglobin) and the same kind of defect in the genes that underlies thalassaemia underlies many other genetic diseases like cystic fibrosis and muscular dystrophy. It was the study of thalassaemia that changed Weatherall from being exclusively a haematologist and led him into the much wider field of genetics. His experience gives an insight into how medical genetics has developed over the last twenty years and how it has crucially depended on a series of scientific discoveries.

Weatherall's interest in genetics, would have surprised him as a young man, as he explains:

When I started out in medicine everyone thought that genetics was a factor in only a small number of rare diseases, but it's turned out that it has major implications right across medicine – even cancer.

As with many Vane and Iversen, Weatherall was influenced by some of his early mentors, of whom one was Dr Cyril Clark, Professor of Medicine at Liverpool University. Clark was both a physician and a butterfly geneticist; he was interested in applying what he had learnt from butterflies to humans. Like a number of British scientists, Weatherall spent some formative years in the USA where he joined a team of genetics researchers at Johns Hopkins University.

It was at Johns Hopkins that he was able to research thalassaemia – a disease that he had seen at first hand while a medical officer on National Service in Malaya, a region where the disease was later found to be common. He was looking after a children's ward where there was a small child suffering from this disease. Children affected with the disease were kept alive by continual blood transfusions, but in severe cases they became highly anaemic and died.

Weatherall spent about five years in the USA researching haemoglobin (in order to understand thalassaemia) but did not get as far as he would have liked. He says:

> When we started, there were ideas around that the production of haemoglobin was defective in thalassaemia. So, first we had to work out methods for measuring haemoglobin production in a test tube. This we managed to do. The next problem was how could a genetic defect cause a reduced output of a gene product. It was an interesting problem because it seemed that if we could understand what was wrong in thalassaemia, then we might understand what was wrong in many genetic diseases where there was defective production of a particular protein.

Weatherall adds that until they could get at the DNA they had to make models based on bacterial genetics, but these all turned out to be wrong.

Unscrambling the whole chemistry of haemoglobin production took about fifteen years, with four important steps along the way. The first stage was in the mid-1960s when it became possible to measure the way haemoglobin was made. In Cambridge Max Perutz had recently discovered that haemoglobin was constructed out of two chemical chains – alpha and beta – and that both had to be able to communicate. This knowledge made it possible for Weatherall and others to measure haemoglobin production. The second stage, says Weatherall, 'occurred when we could look at the protein synthesis', but this still left out the impact of the genes on the proteins.

It was the third stage – getting at the messenger RNA (which works with the DNA to synthesise proteins) – which enabled Weatherall to come nearer to the secret of haemoglobin production. It was these messengers which played a crucial part in the manufacture of the protein for the blood cells. This third stage was made possible by advances in molecular technology in the UK and the USA. Until those advances were made Weatherall's work was stalled.

The fourth stage, in the late 1970s, was working with the DNA itself – isolating, cloning and sequencing the genes responsible for haemoglobin production. Weatherall's team and several others found about 200 possible defects in the globin gene. These relate not only to thalassaemia but to all the other haemoglobin-based diseases.

An important lesson from this history, says Weatherall, is that much of it was based on other scientists' 'curiosity driven research'. He continues:

> Take the example of fractionating DNA into small pieces. This was pioneered by two American Nobel prizewinners, Nathan and Smith. They were studying viruses, not haemoglobin. They wanted to know why the growth of some viruses was restricted in certain bacteria. It turned out that the bacteria had special enzymes for chopping up DNA. Those enzymes turned out to be of immense value in chopping up human DNA and it made a lot of this haemoglobin research possible. The next stage was

finding out how to sequence the bits of DNA, because DNA has to be in a precise order, and that was discovered by Frederick Sanger in Cambridge. The whole of medical research is based on an accretion of individual advances, often in widely scattered places and frequently totally unrelated.

The practical value of some of this genetic research has already been proven. Parents can be screened for thalassaemia and if a thalassaemic embryo is conceived the pregnancy can be terminated. In two countries where thalassaemia used to be very common – Cyprus and Sardinia – very few children have been born with this disease in recent years as a result of the introduction of these screening techniques.

In Britain this haemoglobin research has contributed to our understanding of cystic fibrosis, muscular dystrophy and haemophilia. Since the haemoglobin genes were sequenced and many different mutations were found in thalassaemia, it has been discovered that the same type of mutations turn up in cystic fibrosis, muscular dystrophy and haemophilia and most other genetic diseases. Thus thalassaemia has given us a reasonable idea, says Weatherall, of 'the repertoire of things that can go wrong with our genes'.

The long-term aim in these case would be 'gene therapy', i.e. replacing the defective gene. The idea may be simple but the technology is extremely complicated, as Weatherall explains:

> The problems will be not so much in isolating a good gene from the appropriate DNA but in transferring it into the appropriate cell. Then the problem is to make sure that it has got the right regulatory bits and pieces so that it functions when it gets inserted into the genome.

The major regulatory regions for globin in one gene were found by scientists in London, and for the other major globin gene by scientists in Weatherall's group in Oxford.

One result of successes like these is that politicians press for science to be more targeted. Yet, as Weatherall says, much of science 'is not ready to be goal directed'. He adds:

> The major advances usually require work in the fundamental sciences which is often done through curiosity rather than with any goal in mind.

Weatherall has been asked from time to time whether it would not be better to concentrate on solving an immediate practical problem, such as finding a cure for backache, rather than spending money on basic research. The answer, says Weatherall, is not as simple as the question.

> If you gave me £2m to solve low backache I could push the research very hard, but I'd soon come up against lots of areas that we still don't know much about, such as the chemistry of the bones and the joints, and that takes you back to basic research. If you haven't done the basic research you won't be in a position to address the practical problems.

He adds somewhat despairingly:

> When you try to explain to politicians and civil servants what science does and how it

works there's a curious kind of switch-off. They say yes as if in agreement, but you know they haven't understood.

Weatherall is but one in a long line of distinguished medical scientists who have been trying to explain how science works and why targeted research does not necessarily achieve the intended results. Medawar makes some of the same points in his autobiography. On becoming Director of the National Institute for Medical Research in 1962, Medawar said, he saw it as his principal duty

> . . . to create and sustain an environment conducive to the advancement of learning and to make recommendations to the MRC for funding areas of biomedical research most urgently in need.
>
> My function was not what one Minister of Science supposed it to be, that of issuing orders from my desk about the research in which each member of the staff would be engaged. I told the Minister when he visited the Institute that if my function had been that of a petty academic dictator, then all the people in the Institute whose services the MRC was most anxious to retain would promptly seek employment elsewhere.

Medical science has done better than many other areas in the fight for funds in recent years, but it has also become a victim of its own success. As one disease becomes treatable so another seems to emerge. At the same time, as our material well-being increases our expectations grow and more is demanded from medicine.

Now, says Weatherall, we are expecting too much, too quickly. A key concern that he has for the future is the training of more people to fill the sort of role that he has, acting as a middle man between basic research and clinical medicine. In the last five years medical practice, he says, has come much closer to basic research, so you need to breed a generation of people who can talk across the two fields 'because the science has become so sophisticated and the benefits so great'.

—— 5 ——
MOLECULAR BIOLOGISTS:
THE BLUEPRINTS OF LIFE

*Almost all aspects of life are engineered at the molecular
level, and without understanding molecules we can
only have a very sketchy understanding of life itself. All
approaches at a higher level are suspect until confirmed
at the molecular level.*
Francis Crick in *What Mad Pursuit.*

Molecular biology has become one of the fastest growing and most successful branches of science. Crick's observation above explains why it has attracted so much attention. Not only does it provide important information for understanding the biological processes in humans, it also gives insights into animal and plant biology, and thus plays a major role in agriculture and pharmaceutical manufacture as well as in medicine.

Crick's interest in this field originated in his collaborative research with James Watson into the structure of DNA (DNA is the acroynm for deoxyribonucleic acid, which contains the blueprint for genetic information). Crick and Watson, together with Maurice Wilkins, were awarded the Nobel prize in 1962 for their discovery that DNA 'is usually found in the form of a double helix, having two distinct chains wound around one another about a common axis'.

The research of Crick, Watson and Wilkins as well as that of Rosalind Franklin, who also contributed to solving the structure of DNA, but who died before the Nobel awards, all built on the work of others stretching back in piecemeal fashion to the 1860s.

Subsequently, others in their turn have pursued research based upon the work of Crick and his contemporaries and the whole field of cell biology has grown enormously in the last thirty years. New techniques of study, new lines of enquiry and new applications, principally in medicine and biotechnology, have combined to drive this field forward. Many questions remain, such as how particular cells recognise each other and how some cells become defective from the body's point of view. The practitioners in this field call themselves either molecular biologists or biochemists, depending on the emphasis of their work, though this distinction has become blurred.

A substantial proportion of this work is funded by the Medical Research Council (MRC). The MRC spends around £50m a year on molecular biology, of which about 20% goes to the Laboratory of Molecular Biology in Cambridge, which has produced eight Nobel prize-winners; one of the Nobel laureates – Fred Sanger – has the unusual distinction of having been awarded the prize twice. The other sources of finance are charities like the Wellcome Foundation and pharmaceutical companies. A relatively small amount of funding also comes from the Science and Engineering Research Council (SERC) and the Agriculture and Food Research Council (AFRC).

Given its past achievements, this field is highly regarded by the Science Research Councils and it is supported better than some other areas of science, but even here scientists feel that limited budgets for new equipment and difficulties in funding new research are stifling their best efforts.

The five scientists whose work is outlined in this chapter represent some of the principal sectors in molecular biology and biochemistry. The first three have been awarded the Nobel prize. They are Dr Max Perutz, who discovered the structure of haemoglobin, Dr Cesar Milstein, who discovered monoclonal antibodies, and Dr Peter Mitchell, who demonstrated how energy is stored and transmitted through cell membranes. Professor Sir Hans Kornberg and Professor Jean Thomas are both working on aspects of molecular transport and cell recognition. Dr Peter Mitchell sadly died while this book was being completed, which is why he appears in the past tense.

One of the characteristics these scientists have in common is that they have not been intimidated by the conventional view. Although scientists like to say that each new generation subverts the theories of their elders it is actually rather rare for scientists to go against the grain of received wisdom. You have to be peculiarly intuitive or firm-minded to take a particular line when everyone around you says you are mad. When Max Perutz began studying the structure of haemoglobin using X-ray crystallography in 1937 he says he was regarded as crazy. (Haemoglobin is the protein molecule that transports oxygen in the blood, picking it up from the lungs and then discharging it into the tissues.)

When Perutz arrived in Cambridge as a young postgraduate student from Vienna in 1936, he says 'proteins were regarded as black boxes'. People had realised, he says, that a knowledge of proteins was essential for the understanding of the workings of living cells but apart from that little was known about proteins. Their structure was a closed book. However, his boss, the legendary J. D. Bernal, had discovered that protein crystals gave X-ray diffraction patterns, and Perutz therefore believed that, in principle, it should be possible to determine the structure of proteins using X-ray crystallography.

The wavelength of X-rays is comparable to the spacing between atoms in many crystals, and when X-rays are shone on to a crystal they are diffracted to produce a pattern on a photographic film which is determined by the arrangement of atoms in the crystal. As Crick explains, in his book *What Mad Pursuit*, crystallography enables an experimenter to determine the positions of all the atoms in a molecule by visualising the density of the electrons that surround each nucleus. But Crick, who worked in the same laboratory as Perutz and saw him develop his crystallographic methods, also points out that crystallography is a difficult technique.

Perutz was strongly influenced by Bernal:

> He inspired me with the belief that X-ray crystallography would one day solve the structure of the molecules of life. He believed that it was the only method of solving this great problem, but at the same time it was generally thought that it would be impossibly difficult because X-ray diffraction hadn't even solved the structure of ordinary sugar which contains only about 40 atoms. The idea that you could solve the structure of molecules in which you have thousands of atoms was regarded as crazy.

However, Perutz was stimulated by the possibilities of success rather than overawed by the likelihood of failure and began the study of the protein molecule haemoglobin using Bernal's X-ray approach. This raised some eyebrows among Perutz's colleagues as he recalls: 'They thought I was mad to take on such a problem.'

The first haemoglobin crystals were made by a Cambridge physiologist, Gilbert Adair, and presented to a grateful Perutz, who remembers the occasion well: 'It looked like sapphire with a beautiful flat surface and a hard edge and it sparkled in the light.' Perutz then took X-ray photographs of it, publishing his first paper on haemoglobin in 1938. But his work had only just started. It was to be another 23 years before he could construct a model of the three-dimensional structure of haemoglobin. In the meantime he struggled for funds to finance his research, aided by the support of Lawrence Bragg who headed the Cavendish Laboratory where Perutz was working. Perutz needed the money because he did not lecture and could not obtain a university post. He was he says a misfit:

> I was a chemist working in the physics department on a biological problem. I couldn't teach physics because I hadn't learnt it, I couldn't teach chemistry because I was in the wrong department and I couldn't teach biology because I wasn't a biologist. So the university wouldn't give me a job.

Eventually, through the influence of Bragg, who himself won a Nobel prize, Perutz gained the support he needed, initially from the Rockefeller Foundation and later from ICI. In 1947 he and John Kendrew were taken on by the Medical Research Council. Until 1962 their unit continued to be housed in the Cavendish Laboratory. In that year it moved into the Laboratory of Molecular Biology which the MRC had built for it.

In pursuing his research Perutz faced endless technical difficulties. 'I tried a great variety of methods to get at the haemoglobin structure and all of them failed until I hit upon the right one sixteen years after I had started.' It then took another six years' work for the structure was solved. Perutz has described the moment of discovery as being similar to falling in love. 'It was the most exciting thing that has ever happened to me, a fantastic thing to see something that nobody has ever seen before.' What he saw was a molecule composed of about 10,000 atoms: mainly nitrogen, oxygen, carbon, hydrogen, and sulphur, with just four atoms of iron. As well as revealing the structure of the haemoglobin molecule, Perutz was to show that it changes its shape very slightly as it picks up oxygen and discharges it into the tissues. Having determined the structure it then took him another seven years to find a way of photographing it. This research, too, was at first immensely frustrating.

Meanwhile Perutz's colleague, John Kendrew, was doing parallel research on myoglobin, the protein that transfers oxygen from the blood to the muscles. Using Perutz's method, Kendrew solved its structure in outline in 1957, and worked out the detailed arrangement of the atoms in 1959; his was the first complete description of a protein structure. When Perutz and his collaborators solved the structure of haemoglobin in outline in the same year, they found that it was made of four nutrients, each closely resembling myoglobin. Perutz and Kendrew shared the Nobel prize for Chemistry in 1962.

These discoveries produced 'a sensation' in the academic world but they also had important implications for medicine. A variety of inherited diseases are related to haemoglobin. As Perutz points out:

> Their nature was obscure until we knew their structure and were able to pinpoint in atomic detail what these diseases were due to. This was the first time ever it was possible to understand the nature of genetic diseases on an atomic basis, and our findings became a model for understanding other diseases.

Sir David Weatherall, Regius Professor of Medicine at Oxford University comments: 'Max's work opened up the whole field of defective structures in haemoglobin so that for example we now understand the basis of sickle cell anaemia.'

In the last few years, as a result of advances in genetics, Perutz's work has born fruit in a new completely unexpected way. He says:

> Now that genetic engineering has become practicable the knowledge of the haemoglobin structure has become useful in a way that we could never have foreseen, because it has made it possible to engineer haemoglobin as a blood substitute for transfusions. Some biotechnology companies are going to bring various genetically engineered haemoglobins to market which will be useful – they will be free of viral infections.

Perutz has been dismayed by the way science and scientific research has been regarded in the last decade. He says that it is a tragedy that parliament and government and a great deal of British industry, too, 'is unaware of science and what it can do'. He adds:

It was a tragedy that Mrs Thatcher, who studied science at university, as Prime Minister thought that science and manufacturing were unnecessary, and that we were living in a post-industrial age. This was the greatest piece of nonsense ever pronounced, and it has penetrated government and determined government policy on science.

Like many other eminent scientists, Perutz believes that great damage was done to British science during the 1980s, and that without a substantial and immediate improvement in funding our scientific capability and therefore our economic prosperity is endangered. In talking of science policy he includes engineering, lamenting its poor status. He says:

While Germany and Japan have built great engineering schools which have made them dominant in the world, in Britain engineering has been regarded as hardly an academic subject. It is the chief reason why our economy is in such a parlous state.

In some areas of molecular biology it has taken decades for the potential of the original research to be fully realised. A classic example of this is the case of monoclonal antibodies, which are a line of antibodies manufactured by a single cell. This field was pioneered by Dr Cesar Milstein, whose research, and that of his colleagues, has enabled today's biotechnology companies to clone antibodies for use in new drugs.

Milstein first went to Cambridge in 1958 as a postgraduate student from Buenos Aires to pursue his interest in enzymes. In 1961 he returned to Argentina. Two years later he came back to the MRC's Laboratory of Molecular Biology and began his research on antibodies. What particularly enthralled him was the sheer diversity of antibodies in the human body. How was it possible for the body to make so many millions of them? As he explains:

It was a major mystery in immunology. Each of us makes a large number of antibody molecules in our life and having made one we remember how to make others so that when we are ill we can recall the information as if it was information stored in a computer. To try and understand that was a big challenge.

Milstein and his team began their researches by looking at the chemistry of the proteins in the antibodies which gave them the basic outlines of the problem. The next stage was to look at antibodies using genetics. Milstein proposed the theory that the antibodies were made as a result of a 'high rate of mutation in the DNA where the antibodies were expressed'. If this proved to be the case then 'it would explain the genetic mechanisms involved in making antibodies and it would explain the origin of antibody diversity.' To test this theory Milstein wanted to take living cells and put them in a tissue culture, in order to watch the cells grow and make antibodies and then to ask another question: would the descendants of the cells all make the same antibodies? The answer to that would indicate whether the cells were capable of mutating. The only problem with this approach, as

Milstein was all too aware, was that if you take a cell from a body it will die. So instead he used cancer cells because they have the acquired capacity to divide in a culture. To his disappointment they did not mutate as expected. He then tried some other cancer cells he thought were more suitable, but again the experiment failed. As Milstein comments: 'It is not enough to have a good idea in scientific research. What is often equally important is having the right technology to enable you to undertake the experiments.'

For the moment Milstein was completely stuck. But he and his colleagues were also engaged in another piece of research which involved fusing cancer cells. Here the question was: would two or more cells, when fused together, produce two different antibodies or a hybrid? The answer was that they produced a hybrid expressing the genes of both the original cells. This meant that the genes involved in the synthesis of the antibodies were co-dominant – both were expressed at the same time. This result gave Milstein an idea for his other line of research:

> We conceived the idea of doing our original experiment by deriving a new type of cell by fusing one normal cell with a cancer cell. We conjectured that if the cancer cell is making some sort of antibody maybe it will immortalise the antibody of the cell that we can't grow, i.e. the non-cancer cell.

This thesis proved correct. By fusing a healthy cell that was making antibodies with a cancer cell a new cell was formed that was able to reproduce unlimited antibodies of the healthy cell.

Having achieved a successful result, Milstein and his colleagues then ran into the sort of problem that every scientist dreads: when they tried to repeat the experiment it did not work. As Milstein explains:

> The first time we did the experiment we succeeded because we had the right conditions, we prepared it in the correct time and we did it all in the right way. But at the time we didn't know that. As we repeated the experiment we had to solve a series of difficulties until we had found how the first experiment had worked.

This first breakthrough was followed up with a series of other experiments investigating hybrid cells, and these later experiments proved conclusively that each set of antibodies from a particular cell will be identical and will be targeted at a specific infection. The discovery of monoclonal antibodies provided medicine with a medical tool by enabling clinicians to use antibodies as an element of diagnostics. As Millstein says: 'The implications of our work went far beyond our original intentions.' Unhappily, the National Research and Development Corporation (NRDC) which was at that time responsible for the commercial patenting of government-funded research did not consider that Milstein's discoveries had any immediate practical applications. (The blame for this failure has often been wrongly placed upon the MRC.) This case is frequently cited as one of several examples of Britain's failure to exploit financially a scientific discovery. However, subsequent work in this area has been patented for use in the pharmaceutical and biotechnology industries.

This research took many years – seven in all – and Milstein is still developing the field. Like Perutz, he has been able to discuss his work with colleagues working in other sectors of molecular biology in the same building as himself. He also points out that his research rests to a considerable extent on the earlier work of others. He was also helped by four postdoctoral students who worked with him at various stages: Richard Cotton, David Secher, Giovanni Galfre and Georges Köhler. Köhler, who came to the laboratory from Basle, shared the Nobel prize for Physiology or Medicine with Milstein (and also with Neils Jerne) in 1984.

Looking back, Milstein says that many of the research results might now seem obvious, but adds that what looks logical today was imaginative yesterday:

> Whatever you do, it always looks obvious afterwards but at the time you need intuition and imagination to make the leap forwards. Logic and imagination go hand in hand. Logic without imagination will keep you confined, but equally imagination without logic is useless. What is important is how you structure a problem, which is difficult because you don't have a map of where you are going. So you move in two or three different directions at the same time. I always say, it doesn't matter how crazy an experiment appears to be; if you can do it easily, go ahead, but if it takes a lot of work and if it's crazy then don't do it.

Monoclonal antibodies were an unexpected by-product of basic research which made a direct and significant contribution to the biotechnology industry. Another case where practical applications were clearly derived from basic research is the revolution in the production of insulin for diabetes sufferers which, Milstein points out, resulted from research into bacteria. Sometimes, however, the contribution of fundamental research to technological developments is not so obvious. An example of this, quoted by Milstein, is the automation of biochemical analysis in hospitals. This grew out of the work of three scientists, Spackman, Stein and Moore in 1958. They developed the first automatic amino acid analyser, which they designed in order to determine the amino acid sequence of proteins.

Perutz and Millstein have pursued their work in the context of a large, relatively well-funded laboratory, the Laboratory of Molecular Biology, Cambridge. This is what leading-edge science is often said to demand; but there are a few scientists who have made breakthroughs in quite small centres. One very unusual example was the late Dr Peter Mitchell who won the Nobel prize for his work in a modest laboratory in a small country house in Cornwall. His research centre, the Glynn Research Institute, is on the face of it a most unlikely setting in which to undertake major scientific research – being situated miles from any academic centre and staffed by only a handful of researchers. But Lord Porter, when President of the Royal Society, hailed the Institute as an example to the rest of British science.

Dr Mitchell was an example of a scientist who doggedly refused to accept the conventional view about a particular phenomenon even in the face of some ridicule from colleagues, and who was ultimately proved right. Mitchell is credited with having revolutionised the scientific understanding of how all types of cell from humans, plants and bacteria transmit and utilise energy.

Two American scientists, Virginia Huszagh and Juan Infante of the Institute for Theoretical Biochemistry and Molecular Biology at Ithaca near New York have commented that Mitchell's success

> . . .did not arise from a lucky accident, but resulted from an acute appreciation of the inadequacy of the prevailing paradigm [i.e. the accepted orthodoxy in his field], a critical analysis of the literature and a good deal of reflection and imagination. . . . What allowed Mitchell to propose his ideas was not that he had access to any new data but that his mind worked differently.

Mitchell's mind was that of a philosopher and a scientist. He says:

> As a young scientist I was always interested in the nature of things. How do you know a thing? What makes one thing completely different from another and how can two apparently totally different things be the same?

In his early days as a fledgling Cambridge scientist in the late 1940s and early 1950s, he said he was caught between two factions – the enzymologists, who were interested in enzymes and the part they played in chemical reactions within the cell, and those who were studying cell membranes, i.e. the fine walls that surround the cells. Intellectually he was in between these two; he felt intuitively that the answer to understanding how cells behave was based both on enzymology and on the behaviour of membranes. He therefore did not really belong to either camp within the scientific community and he fell foul of the enzymologists, in particular, whose chief was the boss of the biochemistry department in which he worked. He said: 'I wasn't popular with my boss because I was doing what was supposed to be of no use; I had my own ideas of what was important in research which I stuck to.' He says that it was David Keilin, a professor of physiology, working in the adjoining laboratory who gave him the confidence to pursue his theories.

What nagged away at Mitchell was:

> . . .the possibility of bringing together ideas from classical chemistry which are to do with chemical transformation (the altering of chemicals in cells themselves) and the ideas of chemical transport which relate to the movement of chemicals through space (i.e. across membranes within the cells).

What he was groping towards was an integration of two concepts – metabolism and molecular transport (the transport of molecules into a cell). 'If you could roll these two ideas together it would give you a much deeper understanding of cellular processes in general. This meant trying to find out how a chemical action drives a transport process. That wasn't understood at all.' For about fifteen years he pondered these questions, first at Cambridge and then at Edinburgh University where he was appointed Director of the Chemical Biology Unit in the Department of Zoology.

In trying to understand how chemicals are conveyed through a cell's membrane Mitchell was faced with two existing theories involving two distinct processes. The first was called 'facilitated diffusion' and was based on the assumption that there was something in the membrane itself that enabled the diffusion of a water-soluble substance through the membrane. Since membranes are made of fat they are impermeable to most water-based substances. Therefore a diffusion mechanism appeared to be the most likely means by which a water-based molecule could get through the fatty membrane wall. But there was also thought to be another process called 'active transport'. In this process a substance is forced through the membrane by means of an electro-chemical action. One aspect of this second theory puzzled Mitchell because part of the theory held that during the process 'the molecules move up an electro-chemical gradient'. Mitchell believed that this could not be happening because, 'no substance ever goes up against a prevailing force according to the laws of physics.' His philosophical turn of mind led him to ponder whether, in reality, these two separate processes could be two aspects of a single process. 'It occurred to me that these two processes were different facets of the same phenomena, that is, the active transport process was a kind of facilitated diffusion.'

Mitchell was able to show that this was indeed the case by thinking about the physics of the problem as much as the chemistry. The theory of the electro-chemical gradient remained intact, but as the chemical substance was pushed up its electro-chemical gradient, so protons were being pushed 'downwards', a notion which made more sense in terms of physics.

In chemical terms, what was happening was that the molecule that passed into the membrane from one side was being altered at a half way stage so that what came out the other side was a derivative. The twin process created a chemical called ATP (adenosine triphosphate), which has been described as an 'energy currency' that powers the functions of the cell in plants, animals and, of course, in humans.

Mitchell published his theory in 1961 calling it 'the chemiosmotic theory' of energy transfer. The word was formed from chemistry (indicating chemical change) and osmosis (the process of absorbing liquid, as occurs, for example, when a sugar lump touches the surface of a cup of tea). He described the chemiosmotic process as a kind of electric circuit in which chemical molecules are driven through the cell by a current of protons.

In 1963 Mitchell fell seriously ill with acute gastric ulcers brought on by stress in his Edinburgh department. Medical opinion advised that he should have four-fifths of his stomach removed, but when he was also told that such an operation had a 5% likelihood of failure he resisted. Instead he opted for a complete rest with some minimal 'appropriate medical treatment', and took early retirement. He then bought a small Cornish country house, Glynn House, financed by a legacy from an uncle, and to the surprise of his family turned part of it into a laboratory. By 1965 the house was restored and a charitable institute was set up, the Glynn Research Institute.

But Mitchell's work was not yet complete – indeed in one sense it is still being developed. 'I still wasn't sure if my ideas about the chemiosmotic process would turn out to be right. If I was right it would be excellent and it would persuade biochemists to take the concept of chemical transport seriously.' Mitchell brought in Jennifer Moyle, a long-standing research associate, as an assistant to help him test his theory. Meanwhile scientists all over the world were also engaged in the same process of testing his ideas. His basic thesis was proved correct, though some of its details remain controversial, and he was awarded a Nobel prize in 1978.

Mitchell's work opened up the field of 'chemical transport' in cell science. This, together with the understanding it brought to how ATP is made and used has a number of medical applications. For example, ATP is used by the body to move muscles and to transmit nerve signals and plays an important part in the respiratory mechanism. But the manufacture of ATP in the body is impaired by the ageing process; thus, the development of drugs which address impairments in the manufacture of ATP, which is a distinct possibility, might help in the treatment of a wide number of diseases. At least one Japanese drug company and several British pharmaceutical companies have worked on some of the ideas that have come out of Mitchell's research. Mitchell was, however, disappointed, like some other British scientists, to find that British companies were far less ready to invest in new scientific advances than were their Japanese counterparts.

Although the Glynn Research Institute is very successful in competing for research support from public agencies such as the MRC and the SERC, Dr Mitchell expressed concern that it was becoming harder to obtain funding for really novel areas of research – a trend that has been noted by many other scientists. Mitchell cited his Research Director, Dr Peter Rich (now Chairman of the Glynn Research Foundation), whose research proposals have in several instances not been funded even though they were rated by peeer review to be of the highest level. One project involved a specific area of chemistry that, as Mitchell put it, 'hardly anyone else was looking at.' The project was recommended by all the referees to whom it was submitted, but it was rejected by a committee that subsequently assessed it. Mitchell argued that the present system of awarding grants results in arbitrary decisions, with fashion often dictating where the money goes.

One of Dr Rich's earlier financial supporters, while he was researching at Cambridge University, was Dr Donald Braben, who was at the time in charge of BP Ventures Research. Braben makes the point that, when money is tight, the peer review system works against innovative science, because no one dares to put money into projects that are seen as somewhat unconventional. He scorns the notion of committees adjudicating on the value of research, saying: 'Committees can only work by consensus, but science is not a consensus operation.' As well as the pressure of the peer review system there is also a drift towards the the idea of 'managing' science. Trying to manage science against some kind of objective financial or practical criteria is a nonsense, according to Mitchell: 'First rate innovative science cannot be managed except by the scientists who are doing the work. If you try to manage it by pushing it around, then you will waste your time.' The best science, he claimed, has not come about by setting scientists targets, because that is simply not the way science works.

One of the growing areas of biochemistry, where the applications seem boundless, is concerned with molecular recognition, i.e. understanding how molecules interact with each other: drugs with proteins, hormones with receptors, one protein with another, proteins with DNA etc. The latter area forms part of the research of Jean Thomas, Professor of Macromolecular Biochemistry at Cambridge University.

Professor Thomas started as a chemist in the 1960s but was attracted to structural molecular biology by the work of Sanger, and of Perutz and Kendrew on protein structure, and by the solution of the structure of DNA by Watson and Crick a few years earlier:

> I became aware of this pioneering work when I was a chemistry graduate student working on small peptides in the 1960s, and I became fascinated with the structure, chemistry and function of macromolecules – the very large molecules, such as DNA and proteins. What interested me in particular was the specificity of the interactions between these complex biological molecules and other molecules, large and small (what is now called 'molecular recognition'), and the way in which these interactions can be precisely explained by knowing the structures of the molecules.

She explains:

> In a biological system enormous numbers of molecules may be constantly colliding with each other. It is vital for the orderly functioning of the cell that molecules are able to recognise the 'right' partners, with which they have a productive relationship, and to discriminate against the wrong ones. Enzymes recognise their substrates (peptide hormones like insulin, for example), neurotransmitters their particular receptors on the surfaces of target cells, gene activating proteins their binding sites on DNA etc.

She adds:

> We need to understand the basis of this recognition, which underlies the whole of biology, in order fully to understand how biochemical reactions are controlled in the cell, how loss of control may lead in some instances to disease, and how, for example, precisely designed drugs might be used to bind to a particular macromolecule. A

study of molecular recognition is not only immensely intellectually challenging, it also has wide-ranging practical applications.

Professor Thomas's particular interest for the past 20 years or so has been in the interaction of proteins with DNA in the cell nucleus. DNA carries all the genetic information, only some of which is read out in any particular type of cell. She explains:

One challenge has been to understand how the DNA is packaged: for example, how over two metres of DNA in a human cell is packed into a nucleus about 1/200th of the diameter of a full stop on the printed page. We, and others, have studied this to find out just how this is achieved. Essentially all of the DNA is wound around a succession of 'beads' which are complexes of basic proteins (called histones), giving a 'string-of-beads' appearance in the (electron) microscope. The string of beads is then coiled into a thicker filament. This complex of proteins and DNA is called chromatin, and is the stuff of chromosomes in all eukaryotic organisms – fungi, plants and animals. It represents a universal means of packaging the DNA in all these organisms.

The study of chromosome structure and the way in which genes are controlled has become an international activity. There is vast scope for increasing our knowledge. She explains:

A large number of groups are working on different aspects of the problem and are using different systems, for example, from yeast, flies, frogs and human cells, to try to define exactly the combination of proteins needed to allow the natural control system for a gene to be reconstituted from the purified component parts in the test tube.

Professor Thomas explains that lessons learned in one system are often useful in another. She says: 'We expect that many of the principles that emerge will be universal, although some features will be unique to particular systems.'

However, knowing what all the components are is only the beginning. She says:

The next challenge is to determine the structures of all of the proteins that assemble at the beginning of the gene. Acquiring this structural information is a necessary step to understanding the molecular recognition between these proteins and DNA, or between proteins and other proteins, that allows the specific process of gene transcriptions to proceed efficiently in the cell.

The molecules Dr Thomas studies are mainly protein molecules and, like Perutz, her principle tool is X-ray crystallography:

The surface of proteins is disordered. In a sense they are all different with no recognisable repeating pattern in the proteins, and it is that that gives them their unique properties. You can't look at structure unless you generate order, unless you have a repeating array of molecules. You can get that order by crystallising them. When you see the proteins in crystal form you can see the structure properly and you see how beautiful they are. The shapes are quite fascinating.

When you have the crystals of protein, you will have a row of protein molecules in three dimensions and then you can shine X-rays through them and the structure reinforces itself.

But this method of unravelling molecular structures is not at all easy:

You can only do so much with crystals – it's still an art rather than a science. You try your best and if you're lucky you get crystals. If you're not then you can't use the X-ray process. A lot of research is to do with getting the techniques of the experiment right in the first place. Fortunately, there is now an alternative to X-ray crystallography for determining protein structures. This is called nuclear magnetic resonance (NMR) spectroscopy, and it can be applied to proteins in solution.

Although the solution of protein structures generally is of tremendous intrinsic interest, it is also of biomedical importance. Professor Thomas explains:

We should remember that this is the age of designer molecules – designed to fit with partners whose structures have been determined at the molecular level. This exploitation of *molecular recognition* has in many cases replaced the tedious and time-consuming screening of perhaps thousands of compounds for biological activity.

There are people in government who might ask why we need to devote resources to this field when other, richer countries are able to put much more money into it. Professor Thomas responds by saying that Britain's record in a number of fields related to molecular recognition, as well as in areas such as chromosome structure and the control of gene expression, is excellent. She adds:

We need to ensure that it stays that way and that, as far as possible, our position is not eroded further. In this as in other areas of science there is a great deal of collaboration as well as competition. It would be exceedingly short-sighted not to recognise that this key area of fundamental research, which abounds in intellectual challenges, also underpins advances in medicine and biotechnology and carries the key to success in the development of new products for the pharmaceutical and agrochemical industries.

Some of Professor Thomas's concerns are shared by Sir Hans Kornberg, Sir William Dunn Professor of Biochemistry at Cambridge. One problem he sees is that the scientific profession is being asked to take on more teaching without being given more human resources to do it. Kornberg, however, is less critical of the peer review system than some other scientists. He has seen for himself, as an adviser to BP Ventures Research, the more time-consuming method of assessing research ideas used by Dr Braben. He says: 'BP Ventures Research supported some very good science which would not have happened otherwise, but its approach to judging the worthwhileness of projects – involving face to face interviews – was expensive.' Pointing to a pile of research projects on his desk that he had been asked to review, he said: 'I receive that many each month. I'd never have the time to talk to each person,' adding, 'It is not the peer review system that is at fault but the fact that funding has not kept pace with the growth of science.'

Like Professor Thomas, Kornberg is interested in molecular recognition, but whereas Professor Thomas researches into non-bacterial organisms, he is looking at molecular recognition among bacteria. Underlying his research is a long-term interest in metabolism. His expertise in this field was first developed when, as a young student, he carried out research for one of the pioneers in the field of metabolism, Hans Krebs. Krebs showed how in most oxygen-consuming organisms food is used to produce energy. This process involves the 'burning' (i.e. oxidation) of substances containing several carbon atoms, linked together with hydrogen, oxygen and other atoms. In this process, oxygen combines with hydrogen to form water, and the carbons are converted to carbon dioxide; at the same time, the energy released in the 'burning' is made available to the cell.

Kornberg's most important work has been to show how fats can be converted into sugars. Many plant seeds, such as those from cucumbers or melons, go through a cycle in which their fat is broken down to make sugar. It is similar to the Krebs cycle but uses more carbon atoms and involves fewer steps. In the Krebs cycle the oxidation involves a compound of only two carbon atoms. Kornberg says that this caused him to wonder how bacteria were able to grow, i.e. how they make all manner of complex cell components 'using a two-carbon compound as their sole source of carbon'.

The resolution of this problem was expressed in what Kornberg calls the 'glyoxylate cycle'; this is based on the synthesis of two molecules which both contain two carbons. Once this process had become clear, says Kornberg, it explained how castor oil converts to sugars in the germinating castor bean, which up till then had been a puzzle. However, it took many months of wrestling with the problem before he arrived at his answer. Indeed it was Kornberg's late wife who helped him to see both the similarity and the difference between the Krebs cycle and what he was observing in a germinating castor bean. She asked him to describe how one saw a Krebs cycle. As he did so he realised that all the compounds that play a part in that cycle are also found as intermediates in the glyoxylate cycle, but the recognition of compounds is not the same as the recognition of the cycles in which they are transformed.

Kornberg's discovery led him to consider other questions:

> I thought what is interesting is not so much what happens inside the cell but how does a molecule get from the outside of the cell to the inside? How does a cell maintain its integrity and yet also admit molecules inside, and therefore how does it know which molecules to keep out?

Working with proteins and sugars in bacteria, he has seen that a cell membrane acts like a medieval city wall which has a number of gateways or checkpoints. It is he says like looking at the old Berlin checkpoints between the Western and Russian zones – 'you're not let in if you're not right.' Some of his recent work has involved examining how sugars gain access to bacterial cells. Not only are the sugars not admitted if they are not of the right kind but if they are admitted they are allowed in only on a very controlled basis, 'so that the cell is neither overwhelmed or starved.' This research is linked to the 'chemiosmotic generation of energy,' mentioned earlier in the section on the work of Peter Mitchell.

In describing the attractions of this research, Kornberg quotes a former teacher of his, Tom Stevens, who likened research to doing a crossword puzzle with half the clues missing and with the other half wrongly numbered. He adds:

> I've always loved doing crosswords and this work is like a gigantic puzzle. Pasteur described his research as seeing what other people had seen but being the first to notice it, and then suddenly seeing it all come together. The satisfaction of this kind of work is that you see things in a context that was denied to you five seconds ago, and then your insight transforms everything.

His current work on fructose uses both biochemistry and genetics to understand how this kind of sugar gets into bacteria. Having first isolated the protein that acts as the checkpoint in the membrane wall for the passage of fructose, he and his team went on to identify the gene that specifies the protein. The next stage was to sequence the gene into its various parts in order to understand more about the protein. Even at this point, Kornberg says that 'we are still at a superficial level'. He adds:

> We can't know the mechanisms until we know the full structure of the protein and for that you need X-ray analysis. We want to get closer to an understanding of what is in biochemical terms a completely baffling phenomenon, i.e. you have a protein that actually shifts something from one side of the cell to the other and then lets it go. It is quite remarkable. But how can a protein act as a pump without chemically changing at all? That is mysterious.

The applications of this work lie in several fields, such as pharmaceuticals and food and drink manufacture. In the fermentation process of drug manufacture one problem that often occurs is that the organism making the antibiotic takes in too much sugar and stops making the antibiotic. The research done by Kornberg's team has indicated at least one way in which an organism can be prevented from ingesting too much sugar.

Biochemistry, like all science subjects, has faced funding difficulties in the last ten years. There are several reasons why Sir Hans Kornberg believes that Britain should maintain a strong research effort in this area:

> If we can understand the behaviour of cells when they are healthy then they can provide clues when the cells go wrong. We can't understand the nature of disease unless we know that. There is an awful lot at the molecular level that we don't know. We don't understand the element of the time function in molecular systems. These are questions that don't just relate to medicine but also to crops and plants. One wants to be able to make two blades of grass grow where only one grew before and help to fight diseases of plants.

Yet, in Kornberg's view, simply putting more money into science is not the sole answer. He agrees with Braben that the very competitiveness of international science is remorselessly pushing up costs all the time. He explains: 'Because science has become so competitive, you have to run like hell, and if you have a bigger group you can run faster. But there comes a point where the finances will not enable you to grow indefinitely. 'Are we therefore trying to do too much research? No, says Kornberg. 'What is wrong is that the way we are doing it is not properly focused, but we haven't discovered a mechanism by which we can focus it.' However, to some extent resources are being concentrated already, but not in the context of an agreed and clearly thought-out framework. The concentration of research grants within dozen or so universities is proving extremely controversial. This is an issue that is discussed more fully in the concluding chapters.

6

ENVIRONMENTAL SCIENTISTS: MAN AND NATURE

The task of the technologist today is not only to feed people with the things they want but to guard and manage the planet at the same time.
Lord Porter, addressing The Royal Society as retiring
President in 1990.

In the popular conception, scientists are often seen as a threat to the environment. But, although science has been responsible, indirectly, and usually under pressure from commercial, social and political interests, for damaging the environment, it has also taken the lead in identifying and quantifying that damage. Moreover, it has an indispensable part to play if we are to redress the environmental depredations of the past. Lord Porter pointed out in his address to fellow scientists that the discovery of ozone depletion was based entirely on scientific research; research which was funded by Britain's Natural Environmental Research Council. The early work on river pollution and on pesticides dangerous to wildlife was also carried out by scientists. Indeed scientists claim that some of their early warnings about pollution were disregarded, because the alternatives to using certain chemicals were too expensive and because the managers of the chemical companies put pressure on governments not to take action.

In Britain today there are several thousand scientists working in the fields of agriculture, forestry, wildlife, animal population, ecology, air pollution and oceanography. They are actively contributing to environmental conservation.

Some of this research is pure, fundamental science with long-term goals, usually funded by one of two Research Councils: the Agriculture and Food Reseach Council (AFRC) or the National Environmental Research Council (NERC). Where the research is contracted for a more specific aim the funds can come from a wide variety of sources: a government department, a local authority, a business such as ICI, an environmental pressure group or naturalist society, the European Commission or a foreign government. For example, British expertise in marine biology and oceanography is often bought by foreign governments.

The combined budgets of the AFRC and the NERC were about £320m in

1991/1992. The AFRC's funding from the government's Science Budget and from the Ministry of Agriculture, Fisheries and Food (MAFF) was £132.2m, to which was added £19m from other sources, i.e. industry and government departments. The NERC's income was about £170m, of which £124.35m came from the Science Budget with the remaining £45m coming from external sources. These two bodies fund an extraordinarily diverse array of research in fields that include soil science, ecology, animal and plant genetics, viruses among animals and plants, river pollution, climatic change, geological and oceanographic surveys and food science.

Both bodies have been under enormous financial and re-organisational pressure in the last ten years, despite their obvious social importance. The Institute of Food Science, which is part of the AFRC, has seen its centres reduced in number, even though thay have performed an important function as sources of independent research into farm animal husbandry and food manufacture. (The government's argument was that such research was 'near-market', and should therefore be paid for by industry. This fitted in with the government's declared view that industries – whether they were financial services or farming – should be self-regulating, which itself was highly questionable. In addition it ignored the fact that, left to itself, the farming industry and to an extent the food industry too, would not start funding research formerly undertaken by government agencies.) The AFRC as a whole has seen its staff reduced from 6,300 in 1983/84 to 3,600 in 1991/92. Since 1988 the number of AFRC sites has been cut from 24 to 11. Commenting on these cuts, the AFRC Director-General, Tom Blundell (who was appointed in 1990), says: 'No one can say that British science is level funded when faced with those sort of statistics.' (The government has argued that science spending has kept broadly in line with Britain's Gross Domestic Product (GDP)).

The NERC has seen proportionately fewer cuts, but it has had to bring in an increasing amount of revenue – now about 25% – from external sources. There have been two consequences. One is that scientists are having to become salesmen, a role for which many of them are ill-fitted and one which detracts from their research efforts. The second is that there is much less money for fundamental research. The £45m that came in from industry and from individual government departments in 1991/1992 was not for fundamental long-term research, which is the basis of all science. It was usually for two- or three-year projects which had very specific, highly targeted objectives. In studying natural phenomena, which is what all these scientists are trained to do, changes take place very slowly. Five years is but a micro-second in the life of a species.

The problems facing the AFRC institutes are set out by Trevor Lewis, Director of the Rothamsted Experimental Station, one of the oldest and most respected agricultural research centres:

> The severe under-indexation of funds means continual erosion of resources resulting in staff cuts and fewer long-term projects. [Total staffing across all three sites

comprising the Institute of Arable Crops Research has fallen from 800 ten years ago to 430 today.] So we try to make good the gap in funding by seeking soft money, which is invariably short-term. The decline in resources has become very serious, because everyone is spending more time chasing funds and less time doing research: I have about 90 project leaders and they are all chasing for funds from within Britain or Continental Europe. The whole research effort will wind down if we are not careful.

Lewis points to three notable achievements of the last twenty years at Rothamsted: the development of pyrethroid insecticides, research on nitrates and research into the mechanism of resistance to insecticides. These have all been long-term projects. The synthetic pyrethroid insecticides took nearly twenty years to develop and were launched in the 1970s. Today, world sales of this group of insecticides exceed £1bn, with royalties going to the British Technology Group and the Treasury. The chief advantages of these compounds, says Lewis, are their low toxicity and non-persistence, i.e. they do not remain active in the soil for a long time after use. Twice during the twenty-year research period the managers of Rothamsted were advised to halt the research because they were told that it would never get anywhere. Fortunately, says Lewis, the advice was ignored. Today, he wonders whether such a long-term piece of research would ever be approved.

Since the mid-1980s, the emphasis, particularly of Ministry of Agriculture-funded research, has been on short-term projects; to such an extent that research, Lewis claims, 'is now driven as much by policy issues as by good science'. An example, he says, is research into alternatives to straw-burning, which is now in the course of being banned. The question is, what effect will this have on the soil if the straw is ploughed in? Should there be exemptions? The simple answer to the first question is that straw ploughed in would increase the level of nitrates. The answer to the exemption issue is more difficult. But the issues are really rather complex, so definitive answers cannot be given over a short time-scale. 'We won't really know the impact that the legislation will have for ten years.' He adds: 'Ministers want quick answers, but nature works with long time-scales.'

The lack of commitment to long-term research – i.e. ten years at least – particularly by the Ministry of Agriculture, has become a real problem. It affects both the quality of the science that is done and the way that scientists are now managed. Lewis explains: 'The opportunity for a brilliant individual to see a niche and pursue it without worrying whether he'll receive the resources has been much reduced.' The present three-yearly research reviews and quarterly reporting procedures imposed upon the agricultural institutes severely limit curiosity-driven research. Lewis poses the question:

How do you compare one person who spends twenty years on one piece of research and comes up with a brilliant idea and another person who does focused research and meets tight objectives every three years? You can't. They are not comparable.

Long-term research on nitrates has been better protected from cuts in funding than some other work because of its obvious environmental and health import-ance. Among the Rothamsted scientists investigating nitrates are David Jenkin-son and David Powlson. One problem they have looked at is 'nitrate leaching', i.e. the draining away of nitrates – from the use of fertilisers – into rivers and underground aquifers. By undertaking a series of experiments on the Rothamsted estate, Jenkinson and Powlson discovered that nitrate leaching could be sharply reduced if nitrate-based fertilisers were not used in the autumn. (If they are applied in the autumn there are no crops to hold the nitrate in the soil, whereas the converse is true in the spring and summer.) The autumn use of nitrates has largely ceased, says Jenkinson, as a result of the Rothamsted studies.

However, they and other scientists made another discovery, that the dominant contribution to nitrate leaching comes, not from fertilisers, but from 'the mineralisation of organic matter already in the soil'. This is especially true in cereal production. There appear to be several reasons for the high levels of nitrates found today in some agricultural soils, says Powlson, and scientists are still researching into the relative importance of these causes. One is the growing intensiveness of arable farming since the 1950s, another is the presence of more chemically combined nitrogen (mainly ammonia and nitrate) in the atmosphere, while a third, and by far the most surprising to non-scientists, is a legacy of the ploughing up of grasslands for cultivation in the 1940s and 1950s. The food shortages of those years encouraged the ploughing up of pasture; one result was that the decomposing grasses led to an increase in nitrates in the soil. Powlson says that it is possible that as much as 50% of nitrate leaching is a result of converting grasslands to cereal production forty to fifty years ago. Research on this still continues.

Most people assume that agriculture will be affected by global warming but it now appears that the natural carbon in agricultural soil will itself affect atmos-pheric temperatures. This has been researched by a team of scientists at Reading University and Rothamsted. One of the Rothamsted reseachers is David Jenkinson.

Jenkinson explains the importance of this research:

> There is about twice as much carbon in the top metre of the world's soil as there is in the atmosphere. We therefore wanted to know how an increase in temperatures might increase the rate of decomposition of carbon in the soil.

The team of scientists undertook studies in a number of countries – both temperate and sub-tropical – reporting their results in 1991. Their conclusion was that if world temperatures rise by about 0.03 degrees centigrade a year – which is the level regarded as most likely by the Intergovernmental Panel on Climate Change – then the additional release of carbon dioxide from soils would increase the global output of carbon dioxide by about one-fifth over the next sixty years. However, the researchers add that more work remains to be done, including an examination of future patterns of agriculture across the world and of the complex relationship between land use and climatic temperature.

A growing field of agricultural research is in animal and plant genetics. The AFRC is currently supporting teams of researchers at Edinburgh and Nottingham universities; Edinburgh is home to the Centre for Genome Research, which is mapping the genomes of farm animals. Other AFRC-funded centres have also been engaged in genetics research. An interesting example has been the genetics of aphid resistance to insecticides, which has been researched at Rothamsted. During the 1960s a number of insects became steadily more resistant to insecticides. One of these insects was a particular aphid that feeds on the leaves of potatoes and sugar-beet and also on flowers such as chrysanthemums, which it damages by spreading a virus. Hence the need for an insecticide to kill the aphids. Research by scientists at Rothamsted in the early 1970s confirmed that the aphid was resisting the insecticide by producing an enzyme in its body which detoxified the insecticide. This resistance was passed on to succeeding generations since it arose from a mutation in genetic make-up of the aphids.

This breakthrough still left many questions unanswered; for example how did the aphids produce the enzyme and how were they able to survive in the face of higher levels of insecticide usage? How did the genes which regulated the enzymes function? How were the insects able to develop an increased ability to degrade the insecticide? One of the scientists involved in the research, Alan Devonshire, says that there were two possibilities. 'Either they were producing a mutant enzyme which was more efficient at coping with the insecticide or they were just producing lots more of the existing enzyme.' Devonshire and his team discovered that the aphids' production of the existing enzyme was going into overdrive. To his surprise he found that 'the most resistant kind of aphid was directing as much as 1% of its total protein production towards this one enzyme, which is quite a dramatic switch of resources.'

How this was happening was not clear. He explains: 'There was only limited evidence from other research into genes and their relationship to enzymes to show that simply having more copies of the same gene would be exploited to produce more of the same.' Meanwhile he came across evidence from a completely different source that the multiple copying of genes to combat a foreign substance could happen. The source was cancer cells. Elsewhere in the scientific world it had been discovered that cells taken from cancer patients and put into a tissue culture to which drugs were added would produce resistance to the drug. These cancer cells, says Devonshire, were behaving in a way that was similar to the cells of the aphids: they were producing a particular enzyme to overcome the toxic effects of the drug. He therefore inferred that the same genetic process was responsible in both cases, though it took several years before this was proved for aphids by exploiting the standard techniques of molecular biology.

The research has been much quoted within this field and has been shown to have implications for other areas of insect research. For example, an investigation into mosquito resistance to insecticide which was undertaken in parallel to this research showed that their resistance works in the same way as it does in aphids, that is, by 'gene amplification'. One of the implications of this study, says Devonshire, is that organisms are more 'plastic' than we thought, that they have the potential to continue to increase the number of copies of a gene and that this happens much more frequently than the mutating of a gene. It is also giving us a new angle on evolution. We have been schooled to think that when a species survives a new threat it is as a result of random mutations in the genes of some individuals, which makes their descendants 'fitter' for the new circumstances. But it now appears that, in addition, the survival of a species can be the result of its ability to generate more copies of an existing gene during an emergency.

These kind of insights have been important in the whole area of insecticide research, which has blossomed all over the world in the last five years. With the ever growing resistance of pests to chemicals that were designed to thwart them, most countries are increasing their efforts in this kind of research.

Another rapidly expanding field is plant genetics, the discipline which fathered the so-called 'green revolution'. A fairly recently started venture in this area, for example, is a programme of research into tomato ripening under the direction of Professor Don Grierson at Nottingham University. Grierson and his team are investigatinging how tomatoes switch certain genes on and off as they grow and ripen. Some of Grierson's earlier research has suggested that it might be possible to produce tomatoes that can be stored much longer without degrading. Another possibility is that mass-produced tomatoes might be bred that would be capable of ripening more fully and have an enhanced flavour but without losing their firmness.

The use of genetics in agriculture has considerable possibilities. The opportunities are ambitiously underlined by Professor Tom Blundell, though whether he will receive enough funding for this as AFRC Director General is another matter. He says:

> The case for mapping genomes in plants and animals is now very strong, because one wants to identify the genes that are important for certain traits; there might be genes that give natural resistance to disease or have other features that one would be interested in.

He adds:

> If one can identify a particular gene, one can transfer it, either by traditional breeding techniques which are made more efficient if you know where the gene is, or you can do it by genetic engineering. That second route will be dependent for its success on public acceptability.

He cites the much-publicised example of a genetically engineered sheep that makes a drug in its milk. The sheep is perfectly normal 'except that it puts out a very valuable and expensive molecule which one can purify afterwards.'

Blundell's more radical view envisages the agricultural industry becoming the base for a whole range of products, many of which are not currently associated with agriculture, such as chemical and pharmaceutical products and fibres. Genetic engineering of plants would make this perfectly possible. The fact that we do not associate industrial-type products with agriculture is because we have forgotten our own history, Blundell points out: 'A lot of our industries used to be based in the fields: fibres, products for energy, herbal cures, chemicals all came out of agriculture.'

As a specific example of an alternative and genetically engineered use of agriculture he cites rape-seed production:

> One could engineer the rape seed and get a whole lot of different oil-based products for use in plastics, cosmetics, vegetable oils, fuels and polymers. One can target a chemical into a seed and then the seed becomes a very concentrated production unit for specialised bio-chemical materials. We would be using the genetics of the plant and using our knowledge of where the genes are situated to change the plant's control elements in its genes and thereby get the plant to produce a slightly modified chemical.

Using plants in this way would, Blundell argues, 'open up huge new possibilities' and would allow us to have 'a sustainable and diversified agriculture'. It has to be an area in which we can expand, he says, because we have all the chemical and biological resource in our science base to do this, and it would make up for some of the loss in our mainstream manufacturing industry. 'I'm convinced it is an area that will stretch from biotechnology and fermentation to animals and plants, and it will complement medical science.'

However, Blundell knows that his vision is not shared by the policy-makers in government:

> It is very difficult to get the government interested in using agriculture in this way, because it is so obsessed with agricultural surpluses and the failure of manufacturing industry. But the government needs to look forward and to appreciate where we are strong scientifically. In genetics and the plant sciences we are very strong. So we should maximise that strength.

One argument that has been put against this view – in a House of Lords report – has been that gearing our agricultural resources towards industrial objectives would be uneconomic. But Blundell rejects this view. 'The economics are much more favourable than people think.'

While there has been an emphasis in recent years on using our environmental resources in the most efficient way, there has been an equal concern that financial and population pressures have led to large-scale damage to the environment. When local authorities or central government initiate an enquiry into some environmental issue they frequently call on one of the institutes of the NERC. The NERC institutes, in turn, work closely with about twenty universities and have contacts with another twenty. Thus the NERC and the university system represent an impressive source of expertise in the environmental sciences, ranging from freshwater and terrestrial ecology to Arctic and Antarctic studies as well as marine and atmospheric sciences.

One focus for this research has been water and sea pollution together with river management. Some of the leading researchers in this area come from the Institute of Freshwater Ecology's study centres, for example in Ambleside in the Lake District and East Stoke in Dorset.

The East Stoke laboratory was originally set up in 1929 as a charity and subsequently developed into one of Britain's leading freshwater research centres. It claims to have sampled every river in Britain for fish and parasite content. Some of its research, for example on nitrate levels, goes back 25 years. Since Lord Rothschild issued his famous (or in some people's opinion, infamous) report on the customer-contractor principle in the 1970s in which he proposed that government customers should determine at least 25% of the research that is funded by the Research Councils, some of the basic research at this centre has been curtailed. However, the laboratory still maintains wide research interests and its methods and approach are often ingenious.

A current project, which is funded by the EC, is investigating pesticide sedimentation. Samples of river water are 'freeze-dried' so that the sedimentation can be analysed by chemists and microbiologists partly for its nitrate content. Harry Casey estimates that nitrate levels in local Wiltshire and Dorset rivers have doubled over the last 25 years. He and his colleagues have also looked at some of the rivers in East Anglia, which they claim have nitrate levels in excess of EC limits.

One of the centre's long-term projects has been the monitoring of fish populations – for example, the number of salmon in the river Frome – studying fish size and counting numbers of fish entering the river. The study has been carried out with the aid of video cameras. It has shown that the average size of salmon in the river has halved since 1945, from ten kilograms to five kilograms, and that the number caught by fishermen has also fallen. But the reason is not yet clear. As Dr Mike Ladle who has worked on the study says: 'We are not certain whether it is due to changed conditions in the river or the sea.'

Sometimes the sources of environmental problems in our rivers seem – at least at first glance – surprising. Ladle gives an unusual example:

> The effect of watercress farms if they become too intensive can be quite damaging to rivers because these farms can generate higher levels of sediment than the streams and rivers can adequately deal with; too much sediment lowers the fish population because the fish require clear river beds.

Fish populations can also suffer from the deep dredging of rivers, which has been the subject of an investigation led by Dr Clive Pinder at the Monks Wood Experimental Station. Pinder and his colleagues have been studying what is called 'channellisation', i.e. the practice of digging deep, rectangular channels in a number of big rivers such as the Severn, Trent and the Great Ouse. This study has been financed by the National Rivers Authority and is welcome news to anglers who have become concerned by the decline of fish in deeply channelled rivers.

Deep channellisation, says Pinder, reduces the number of weedbeds in which fish can shelter and spawn and it results in a faster flow of water through the river which means that there is a lot of displacement of fish downstream. It also results in greater turbidity in the river which affects the algae, thus inhibiting the plankton on which the young fish feed. If there are fewer fish, such as bream and roach, then certain species of fish-eating birds also suffer. Pinder and his team are researching mainly younger, smaller fish, studying what they eat, in what type of channel, assessing water velocity, the total discharge of water into the sea and sewage levels. One conclusion of this research is that it is not water quality that is an issue in these deep channelled rivers but rather the velocity and turbidity of the water.

This kind of research project needs to be of at least five years duration for it to be of any substantial value, but in the present financial climate obtaining funds to guarantee research for that length of time is becoming more difficult. Pinder's research was initially funded by the Department of the Environment, but now the main funding agent is the National Rivers Authority. Pinder's dilemma is that the NRA 'is not interested in the more fundamental research'. So he and his colleagues are faced with the choice of either stopping the research altogether, and therefore sharply reducing the income of his department which is not really feasible, or continuing the work on a contractual basis but with shorter-term scientific horizons. This is one case among many of scientists who are only able to keep their jobs by dint of contracting out most of their time to commercial organisations. This mode of operation is a legacy of the Thatcher philosophy, which held that scientists ought to be market-driven, that their role should be restricted to providing what the market demanded. In the environmental sphere this is clearly a nonsense, since most commercial organisations are not interested in ten- or even five-year research studies, the results of which in any case are likely to push up their costs. It is somewhat analogous to asking a doctor to make a diagnosis of a patient while insisting that he examine only those organs which the patient's employer thinks relevant to the problem.

One aspect of environmental research is the study of terrestrial ecosystems – looking at the whole complex of interactions between animals and plants in a specific environment. An unusual example is the work of Dr Robert Elms and Dr Thomas at a branch of the Institute of Terrestrial Ecology at Furzebrook in Dorset. One of their longest pieces of research has been concerned with the relationship between butterflies and ants. For, as Robert Elms explains,

> If you want to study butterfly habits you very often have to study ants as well. One-third of the world's butterflies survive because of their relationship with ants. The ants move the seeds of plants around on which the butterflies thrive. There is also in many cases a relationship between the ants and the butterflies, in that the butterflies feed sugar to the ant colony in return for which the ants protect the caterpillars of the butterflies from predators.

Elms and Thomas have been researching a small group of butterflies which have an even closer relationship with ants: their larvae have learnt to mimic ant grubs and so are fed by the ant workers. To add injury to insult, some of the butterfly larvae eat the ant grubs, a fact of which, apparently, the ants remain blissfully unaware. But the butterflies can only establish this relationship with a very specific group of red ant, of which there are eleven varieties in Britain. This means that the butterflies are confined to the habitats preferred by their ant-hosts.

Even this apparently esoteric research can have definite practical benefits. Dr Elms instances an example where some conservationists wanted to protect a particular plant which they knew attracted a species of butterfly. What they had not realised was that if the turf surrounding the plant grew too high, then it would attract a different species of ant from the one that the butterflies needed. This indeed happened and the butterfly population fell. But when the turf was cut the interlopers went away and the more familiar ants returned, together with the butterflies.

Another study of butterflies in the context of an eco-system has been undertaken by a colleague of Dr Elms, Nigel Webb. Webb has been studying the nearby Dorset heathlands which have a number of special characteristics. Within a relatively small patch of this sandy countryside there are warblers, snakes, plants, insects and butterflies that are unique to the heaths. One of these is the silver-studded blue butterfly. If the patches of heathland become too widely separated, or if foreign plants and insects come into the territory, then the butterfly population falls. Spray drift of pesticides from adjoining farmland can also affect the butterflies. This detailed knowledge about the Dorset heathlands has been used to advise BP on their nearby oil-drilling operations. Part of the heathlands which have been disturbed by oil exploration are being reseeded with heather in order to maintain the habitat.

Furzebrook's ecology studies have enabled it to win the consultancy contract to advise the Ministry of Transport on how to restore the natural habitat surrounding the extension of the M3 motorway. This is generally called 'environmental impact assessment,' or EIA. EIA contracts have been a growth area for environmental research scientists throughout Britain, though they have faced strong competition from the obviously more business-oriented environmental consultancies. One growing source of EIA contracts has been the new out-of-town hypermarkets, some of whose building permits have been conditional upon the developers undertaking environmental assessment and habitat restoration.

Environmental Impact Assessment studies have become an important constituent of the Furzebrook centre. Indeed 50% of Furzebrook's income now comes from commissioned work, usually from the private sector. But the need to bring in external funding to supplement the grants from the Research Councils raises two important problems, says Furzebrook's Director, Dr Michael Morris:

> It means that we can't carry even the best scientists unless they bring in commissioned research, but a really good scientist may not be suited to be a salesman. . .
> The other problem is that if we do too much contracted work the quality of our science – the ability to be innovative – may fall off.

The conservation studies at Furzebrook are examples of what has become a highly sophisticated national scientific activity, involving mathematical modelling, a computerised national databank and satellite studies of the whole of the British land mass. Statistical analysis plays a significant part in these studies and has given rise to a new subject – biomathematics.

Trevor Lewis at Rothamsted points out that biologists developed some of the early principles of statistics. As he says: 'Agricultural researchers originally led the field in the experimental design and statistical analysis for experiments and this influenced the other sciences.' He goes on to explain why statistics became so important for this particular branch of science: 'How do you set out experiments to give you sensible results when you have hundreds of possible interactions, how do you know that you have not arrived at your result by chance and how do you know the significance of your results?'

The use of statistical techniques and the requirement for very precise data is evident in the butterfly work of Elms and Thomas, who have developed a computerised model to examine the ecology of ants, butterflies and plants. Their models will be used to interpret information that is fed into the national databank of British flora and fauna at the Monks Wood Experimental Station. This databank is being built up from satellite pictures. One of the many uses of this databank is that it can predict bird diversity in a given landscape, as the British Trust for Ornithology recently discovered when it consulted Monks Wood in preparing some maps of Norfolk. The findings were surprising; the satellite photographs showed that a particular area of Norfolk had an extremely diverse landscape which suggested that the bird life would also be similarly varied. Field work showed that there was indeed 'a dramatic correlation between the aerial pictures and first-hand observations'.

The Monks Wood centre was the site for one of the earliest and most celebrated conservation studies in Britain. This study looked at the impact of organo-chloride insecticides on peregrine falcons and other bird species. The research was led by a young biologist and naturalist, Dr Derek Ratcliffe, Chief Scientific Officer at the Nature Conservancy until 1990. He produced what is now regarded as a seminal paper on the effect of organo-chloride insecticides on wildlife.

The first signs that there might be a problem with these insecticides were seen in the late 1950s with massive numbers of birds being found dead on farmland after the seed (which was dressed with insecticide) had been sown. The obvious casualties included seed-eaters, such as sparrows and finches, but it also turned out that predators such as owls, kestrels and sparrowhawks were also decreasing in number.

This latter issue came to light when pigeon-fanciers in South Wales, believing that predatory peregrines were responsible for the loss of their birds, petitioned the Home Office to remove the falcons from their list of protected species. Ratcliffe was asked to investigate. He discovered that the peregrine population was itself falling, but the reason was not clear. He says: 'I suspected pesticides, but the peregrines were nesting far from any fields.'

Ratcliffe went to Scotland, where he had already spent six years classifying the vegetation, and investigated peregrine nests in Perthshire. There he found a nest of eggs that were not hatching. He took them back to the laboratory and discovered that the eggs contained organo-chlorine residues. But that was only the beginning of the quest. As Ratcliffe explains:

> We then had to fit the whole story together. By carrying out a large number of similar studies across Britain we found that there was a timing and geographical pattern that very closely paralleled the use of organo-chlorides.

Ratcliffe then moved on to a related problem – eggshell thinning. He says: 'I'd been puzzled since the 1950s to find that peregrines, golden eagles and sparrowhawks had nests with broken eggs.' So he compared the weights of a series of old peregrine eggs (from naturalists' collections) with current ones and found that the latter weighed less. By making a mathematical model relating egg size to weight he was able to estimate the thickness of the eggshells; he discovered that contemporary peregrine eggs had shells that were 20% thinner than those that were older. He again did more studies, using a series of eggs from collections which had been assembled at different periods, and he noticed that the eggshells started to get thinner after 1946 – the year in which the chemical DDT was first used on farms.

However, these results were not welcomed by the farming community, nor by the Ministry of Agriculture, according to Ratcliffe. While other European countries banned the use of organo-chlorides, the Ministry of Agriculture dragged its feet, finally outlawing their use only when they were banned by the European Commission in 1980. Yet outlawing the use of damaging insecticides does not mean that they will disappear: they can persist in the soil for decades.

When Ratcliffe carried out his study, the Monks Wood centre was part of the Nature Conservancy. Ratcliffe retired two days before this body was broken up and split up into separate organisations covering England, Wales, Scotland and Northern Ireland. The English body is now known as English Nature. About half English Nature's staff of 700 are scientists. Its Chief Scientist is Dr Keith Duff. He says that scientific research has shown that the environment that we take for granted has already changed in quite subtle ways:

> If you look at the chalk grasslands in the South East of England, they are changing in character. They are changing from short springy turf with a great diversity of grasses and plants to one where you have much coarser grasses. The coarser grasses thrive on soils that are richer in nitrogen, ammonia, potassium and phosphates. These elements in the soil are coming from the rain, the air and from pesticides in adjacent farmland.

> As the vegetation changes everything that goes with it changes. Insects change. You lose butterflies because butterflies need short grass, which holds in the sun's heat, for egg laying. Tall grasses have lower temperatures at the base of the stem. If this continues large parts of Kent, Sussex and Hampshire will change the way they look. They won't be the same in general or in detail.

The difficulty, according to Ratcliffe, is to know how to respond. 'Ensuring cleaner air is going to be expensive and managing the land with a much reduced level of extra nutrients is easier said than done.'

However, in one particular area research by scientists at English Nature appears to have been infuential. Studies on the effects of sulphur dioxide emissions from power stations has shown that there has been significant damage to lichens, trees, woodland plants and to lakes. This research has led to a government initiative which will force the electricity generating companies to use a different technology resulting in smaller emissions of sulphur dioxide.

In both agricultural research and conservation studies there is a growing degree of European collaboration. At Monks Wood, the Director, David Hooper, has helped to set up a network of European conservation organisations – participating countries so far include Germany, France, Denmark, Norway, Holland and Finland. The reasoning behind this is partly to encourage an exchange of knowledge and partly financial. By teaming up with scientists from other European countries, Monks Wood can bid for EC-funded projects on conservation, pollution and environmental assessment. Hooper's most recent visit was to the Ukraine where he met a number of scientists from Eastern Europe, a region which is avid for environmental expertise.

This European collaboration also facilitates the exchange of scientists between Britain and Continental countries to work on short-term contracts, say up to two years. A good example is the Long Ashton agricultural research centre. Long Ashton works closely with nearby Bristol university; its principal research areas are plant science, crops and crops and the environment, i.e. the interaction of agriculture and the environment. It also studies hedgerows and headlands. Long Ashton draws a significant proportion of its research staff from mainland Europe, who are all funded by their respective countries.

Long Ashton's director, Peter Shewry, explains the benefits of this collaboration:

> You widen your understanding of your subject by meeting scientists from other countries, and we have benefited a lot from this exchange. Some of our best people have come from the Continent, and they are extremely able. In the rest of Europe many of the brightest people go into science, whereas in Britain we lose a lot to other subjects.

Another reason why it makes sense to set up teams of European researchers is that in the environmental field many of the official initiatives are coming from Brussels or Bonn. Germany has recently passed a law that will make it mandatory for agrochemicals sold in the country to be tested for what it known as vapour phase. Many chemicals vaporise during or after being sprayed and then drift for considerable distances, up to several miles. Long Ashton has begun a study on this problem, funded by the Ministry of Agriculture. This is being carried out in anticipation of the German legislation becoming the model for wider, pan-European regulations sponsored by the Brussels Commission.

As this research and the global warming research at Rothamsted show, an important aspect of environmental science is not only to assess what has happened but to appraise what might occur or what might be legislated for. Another example is the research on coastal erosion at Monks Wood, partly funded by the EC. A particular focus of the research has been saltmarshes. David Hooper says: 'There is a suspicion that some of the saltmarshes are declining, though we are not yet completely sure.' If they are, then it would require the local authorities and the Department of the Environment to spend millions of pounds reinforcing sea defences, or in some more serious cases relocating coastal populations.

There are also a variety of new problems that face environmental scientists. One of these is the synergy of pesticides, i.e. the effects, often unforeseen, when two pesticides are used in combination. Over the last four years a team of chemists, biologists and animal physiologists from Monks Wood, Bristol University and Reading University have been researching this issue. They have discovered that while some pesticides are relatively innocuous when used individually, they are lethal when used together. The particular pesticides the scientists have identified are not used in common a great deal, but when they are their impact can be quite dramatic.

Many of these scientists feel that concern for the environment, despite the fashion for 'green' issues, has not generated the financial support that it deserves. They believe that the environmental dimension that underpins the operation of modern economies is still not understood and that the protection of the environment is not only a cultural, but also economic issue, as Keith Duff explains:

> All of the species in the world are part of an ecosystem. Our whole planet works through a complex set of ecosystems. In Britain, what do we know of that ecosystem? At the general level we do know quite a lot, but at the particular level we don't. For example, we don't know all the invertebrates, all the insects. We don't fully understand their role in soil formation.

He continues:

> Maintaining our environment is important in the final analysis, because if we don't have a healthy environment, the air quality will diminish, the water quality will diminish: these natural systems act as a filter. Everything you do that damages the ecosystem will eventually affect mankind. So much has happened in the last two hundred years that we are now reaping the effects and we will continue to do so for several hundreds of years.

Many of these scientists have already shaped our responses to environmental pressures by submitting their findings to government departments. A significant amount of government legislation in this area has been influenced by scientific advice – examples are hedgerow management, sulphur dioxide emissions and industrial waste – and the government's Green Paper on the Environment drew heavily on scientific representations.

However, these scientists also point out that this scientific expertise has to be based on long-term studies which are properly funded. In order to measure environmental change you have to undertake research that plots that change, year by year, and it needs to be done in detail. So research that is worth anything cannot be instantaneous. But the financial pressures upon these scientists mean that they are now spending more time on short-term applied studies. This detracts from the necessary long-term data-gathering which is essential in this field.

One example of what is happening is evident at the East Stoke Freshwater Centre. This centre now earns 80% of its income from commissioned work with the other 20% coming from the NERC. With commissioned work, the customer dictates the research; the objectives will usually be limited and short-term and they will be intended to address a specific problem. This means that it is not the scientists who judge what is important in scientific terms but the customer whose overriding interests will be financial.

Two things happen in these circumstances. One is that the amount of time available for basic research and fact-finding is sharply reduced and the other is that morale among the researchers falls. Harry Casey comments:

> The base of our scientific information cannot be updated at the level we did before and research into some avenues of freshwater ecology has been stopped. So it is very difficult to operate a high level of enthusiasm in a climate where you are effectively troubleshooting.

David Hooper at Monks Wood makes a similar point when he says:

Although we have been fortunate recently in the high level of outside commissions, it has been achieved at a considerable cost in personal terms to our staff. I do feel that some of my people deserve less pressure in terms of workload, having to meet deadlines and having to negotiate for contracted work.

He adds:

Many of my best scientists – the most original ones – are the least able to negotiate out in the commercial world. So if you force them into a mode where they don't feel comfortable then you put them under a lot of pressure. If that continues it might mean that some of them will leave in order that they can do what they are best fitted for which is to carry out good science. I wish I could allow them a little less pressure, but I haven't the time to market the services of everybody when I have 125 people to support.

At other similar centres concerned with agriculture or conservation the story is much the same, as are the results. Basic science is being squeezed by the need to do short-term fee-earning work in order to make up for the shortfall in government funding. But the answers to many environmental questions lie in fundamental research and not in short-term studies.

7

CHEMISTS:
MANIPULATING NATURE

The statement is frequently made that necessity is the mother
of invention. If this has been the case in the past, I think it is
no longer so in our days . . . We can now foresee, in most
cases, in what direction progress in technology will move,
and, in consequence, the inventor is now in advance of the
wants of his time.
Ludwig Mond, a founder of ICI, addressing the Society
of Chemical Industry in 1889.

Ludwig Mond's forecast of the role of invention and discovery – made with the chemical industry in mind – was very revealing. Like many people of his time, both in Europe and the USA, he believed that the future direction of science was already mapped out. He did not anticipate any revolutionary changes in our knowledge. His was very much an end-of-century view that was shared by a number of people in the USA and Britain. But before we laugh at him with the benefits of hindsight, it is worth pointing out that his views are paralleled today, especially by non-scientists, as we again come to the end of a century. In one respect, moreover, Mond was right. Twentieth-century scientific theory was to run way ahead of its practical application – which does not mean that the applications would not eventually emerge. But, above all, Mond's words still stand as a warning that it is foolish to try to anticipate the direction of scientific research, and therefore misguided to select what research to fund solely on the basis of some presumed market need – a philosophy which has been very much in vogue of late.

The four scientists profiled in this chapter have all worked in very new areas of science. Three of them – Dr Derek Birchall, Professor George Gray and Professor Harold Kroto – have worked in fields that at first appeared to have no immediate practical use.

Gray had difficulty in securing funds because what he was doing seemed to be on the margin of what was then considered important, while Kroto's research was well financed in the early stages but encountered funding problems at a critical stage towards the end. Today both men are internationally recognised for the quality of their scientific research and also for the contribution that their work has made or – in Kroto's case – is probably going to make to industry.

The principal government agency for funding research in chemistry is the Science and Engineering Research Council, but its annual allocation to this subject each year is small. About £9m went into research in chemistry from the SERC in 1991/1992, and the number of good projects that are being turned away is rising. Among those turned down since 1989 have been two Nobel prize winners. They have not been turned down because their research ideas were considered of questionable value. Far from it. In both cases they were told that their proposals were given the highly prized 'alpha rating'. They were refused because there was no money left.

Chemistry now provides the scientific underpinning for a large part of our manufacturing industry. Indeed, the British chemical industry (including pharmaceuticals and oil) is our last remaining industry which is both profitable and large enough to be signifcant on a global scale. Our presence in other industries such as electronics or engineering or motor manufacture has either practically gone or is on the defensive. By contrast, our chemical, pharmaceutical and oil industries are still thriving. They are research-intensive with strong historical links between companies and universities. These links are close, much closer in general than the links between engineering companies and the universities. Not only is there an interchange of ideas between universities and companies but there has long been a movement back and forth between industry and academia in the chemical sphere. ICI, for example, has a number of staff who are on dual appointments, i.e. people who work both for a university and for the company. A number of other chemical and pharmaceutical firms do likewise. The more frequent mode of interchange is for people to make a straight career switch from academia to industry or vice versa.

Professor Stanley Roberts is an example of a chemist who has moved from academia to industry and back into academia again. From a post at Salford University he moved to become a research manager at Glaxo and then returned to an academic career at Exeter. More than two-thirds of his department's research is oriented towards industrial applications of which a common theme is enzymes – natural catalysts used in producing chemical reactions. One of the aims of the department is to help chemical and pharmaceutical companies to use nature's enzymes to re-engineer old products or to make wholly new ones. Thus the processes and the end-products that these enzymes will contribute to will be more environmentally friendly than those they replace.

Professor Roberts puts the case like this:

All living things use enzymes to catalyse natural chemical reactions. They help to split up or put together complex biological molecules, and promote cycles of reactions that control how energy is produced or consumed. Without enzymes, our bodies would grind to a halt and the same is true for plants and animals. Our research is showing how we can use these natural enzymes and allow them to cause chemical reactions which take the place of traditional processes based on toxic metallic catalysis or organic solvents. We have found enzymes that will do things which they have not done before.

These enzymes are derived from animal tissue or from a plant extract or are grown from a micro-organism such as a fungus. For example, the Exeter researchers can take yeast (used in bread-making and brewing) and from it extract certain enzymes that will go ahead and help to make something other than yeast. One application of this research is in the food industry, where enzymes can be used to convert saturated fat into unsaturated fat. An advantage of basing a chemical process on enzymes is that it is likely to work more rapidly than a traditional chemical reaction; for example, a process using a single enzyme might involve only one chemical reaction rather than ten.

A characteristic that makes enzymes attractive for some industrial processes is that they can work at relatively low temperatures. A sophisticated chemical process might require temperatures of 100 degrees centigrade and very high pressures, whereas a large chemical feedstock based on enzymes can function perfectly well at 35 degrees centigrade at amospheric pressure. So there are much lower energy costs.

One important incentive to industry to use enzymes has been the environmental requirement to use fewer solvents, or to replace them with water-based compounds in products such as paints. Enzymes work very well in water. The new chemicals produced from this enzymatic process, says Roberts, are often 'optically pure' as well as being 'chemically pure'.

There are also many interesting prospects for medicine in the use of enzymes to make therapeutic agents and drugs. Researchers have isolated enzymes from bakers' yeast which can be used to make anti-inflammatory steroids. This research derives from the 1960s when it was discovered that steroids in plants could be manipulated to make new steroids to treat inflammation.

The real breakthroughs in this area of science occurred during the late 1970s and the 1980s when it was more generally realised that enzymes could be used to manipulate man-made chemicals as well as those that occur in nature. This has greatly simplified the preparation of a range of new drugs, including some anti-Aids drugs that are now being developed. The new anti-Aids agent, carbovir, is partly based on the research at Exeter. Another area of research is the attempt to synthesise one of the human body's own hormones, prostaglandin. Professor Roberts describes these compounds as 'traffic controllers', catalysing a diversity of reactions in the body from controlling blood clotting to washing the lining of the stomach. The problem in trying to synthesise an artificial prostaglandin is that you want it to behave in a very specific way. Some of this research is quite closely related to the work of Sir John Vane (cited in the chapter on medicine), who made important discoveries about how some of the prostaglandins work; this illustrates how, in any branch of science, the knowledge-base is built up from a variety of sources.

Many scientists would judge that working so closely with industrial partners – there are currently twelve in collaboration with Professor Roberts' department – would cramp their activities, restricting what they can and cannot do. Professor Roberts argues the other way:

> What is attractive about working with companies is that you are talking about real problems. You also find that the area of research may have to shift after three or five years, because the company wants to pursue a new direction in its own research and development, but that is a challenge because it forces you to go into a slightly new area. It means that you are not confined to a single track.

He adds that there is also a two-way flow of ideas and that, in addition, if the department comes up with a breakthrough deriving from its own internally-generated pursuits then a company 'will very often take up the idea and want to develop it'.

There is, he says, a further advantage in having a number of people – about thirty in the case of his department – all working with industrial clients and all possessing a slightly different expertise, in that you create the conditions for intellectual synergy. His team is, therefore, in some senses a multi-disciplinary working group, which by its nature fosters the interchange of ideas upon which science depends. He explains: 'We do have targets for our research, but by having a large team of researchers working on complementary subjects there remains the 'serendipity opportunity' for making new discoveries which is so beloved by scientists.'

In essence Roberts feels that working with industrial clients can enlarge the breadth of scientific enquiry rather than diminish it. Yet as he admits, this satisfaction is partly a reflection of the kind of wide-ranging contractual arrangements that the department has with its clients. He points out that the client relationships are based on three-to-five-year 'research agreements' which are often quite broad in their remit, rather than on 'contract research' which is usually more highly targeted.

Another reason why these agreements work is that the clients themselves often have highly developed research facilities of their own. For example, companies like Glaxo or ICI have research scientists working for them who are intellectually on a par with those in university departments. The research effort in these companies, says Roberts, 'is as good as in some of the best universities'. What this means is that the industrial scientists are in a good position to recognise a good idea from a university when they see it. The two sides are speaking the same language, and therefore a worthwhile dialogue is possible. This view is shared by

Professor Derek Birchall, recently retired Senior Research Associate at ICI and now a Professor at Keele University.

Professor Birchall is the inventor of three ICI products and was one of only four Senior Research Associates at ICI. Like a number of other ICI scientists he held academic appointments alongside his job at ICI. His view on chemical innovation is that it depends on high quality fundamental research both in the universities and in industry. He believes far more in the 'science-push' version of innovation rather than 'the market-pull' version, i.e. that inventions come first and then create their own market. Market demand can produce improvements on existing technologies but it does not stimulate the big inventions. Therefore a company has to retain some nucleus of basic research and it has to encourage its own curiosity-driven research scientists.

Birchall's best known invention is called Saffil. This is a non-toxic inorganic fibrous material which was designed to do most of the things that asbestos does but without being toxic. It is made from an aluminium oxide and it was the first refractory fibre ever made. Its uses have been legion. It has been used in making the US Space Shuttle, in water filtration, in furnace design and in filtering drugs for injection. It has more recently been used in metal-matrix composite materials for the construction of new cars.

The research that led to this invention was, says Birchall, driven by two questions: 'Why is asbestos so toxic?' and 'Could we find a replacement that is not toxic?' The first question is a fairly basic one and is as fundamental as the sort of question that an academic scientist might ask, but its focus is altered slightly by the second, more pragmatic, one. It is the fusion of these two kinds of questions that separates the industrial from the academic scientist. The academic would be unlikely to ask the second question. Yet the solution of both depends upon the availability of the same basic scientific expertise. Indeed, in Birchall's view the possession of a high-level scientific research base both provokes the question and permits the possibility of an answer. 'Without the scientific background here one wouldn't even begin to ask the question and without the knowledge base one wouldn't be able to work towards a solution.'

Another invention is macro-defect-free (MDF) cement. This is a very tough non-porous material as strong as cast iron and ten times stronger than ordinary cement. Again, the invention arose from asking a question, 'Why is cement so friable?' Like a number of other scientists such as Professor Roberts, Birchall has been impressed by natural structures and the way that organisms behave. Birchall had a long-standing interest in aquatic creatures such as crustaceans; he had looked at how molluscs and cuttlefish grew their shells, which he describes as a ceramic made under genetic control. Calcium carbonate crystals are bonded together with a fine layer of protein to give them a strength that is relatively higher than that of cement and which prevents the shell from being crushed by heavy objects. He realised that the shells were made of very dense material with no holes in them, and the absence of holes was one reason for their being so strong. Therefore a stronger cement would have to be equally free of holes. In order to achieve this the cement particles were bonded together with a water-soluble polymer.

This invention was partly created at ICI, with the basic science being done jointly with Oxford University, with both parties engaged in some fundamental research. Again, Birchall argues that to reach interesting inventions you need to have inquiring minds which will examine interesting problems from both an industrial and an academic perspective, and that they should work in tandem.

Ten years ago scientists at ICI began asking another question:'Why are ceramics so weak, why do they have such a low tensile strength?' The answer, the scientists concluded, lay not so much in the ceramics themselves as in the process by which they are made. A new technology was developed which made possible the manufacture of very strong and reliable ceramics. These materials, designed by Dr William Clegg and other scientists, were again based partly upon the natural models provided by sea shells such as mother of pearl. This research also led to what will become virtually a separate business. ICI is now making high temperature ceramic superconductors – materials through which an electric current can flow without resistance. The idea of using superconducting materials to construct electric circuits has been around for some years, but the problem with the known superconducting materials was that they only became superconductive when cooled to very low temperatures by immersion in liquid helium, which is expensive and difficult to use.

However, in 1986 two scientists, Mueller and Bednorz, working for IBM in Zurich, discovered that some superconducting materials could operate at a significantly higher temperature, that of liquid nitrogen, which is much cheaper than liquid helium. This discovery resulted in Nobel prizes for the two scientists and it also set up new aspirations in the computer industry: could these new materials be used in new supercomputers? – for if they could the computers would be far more efficient than those we have now.

When the discovery was announced Birchall says that he and his colleagues recognised immediately that they could make a good ceramic superconducting material. 'Within three months we had made a superconducting ceramic,' he says. In order to test the ceramic the scientists also made a small electric generator which proved to work. ICI is now collaborating with scientists at Birmingham University to produce a range of manufacturable superconducting materials. The venture, which includes the participation of another company, the Cookson Group, is partly backed by a DTI technology-transfer scheme called Link. ICI's intention is eventually to make material and components that will go into a variety of advanced devices.

Birchall reflects that none of this would have been possible if the original questions about ceramics had not been asked all those years ago:

> If we hadn't been in engaged in the research on ceramic materials then we wouldn't now be in a position to make this leap forward. You can't leap on to a discovery if you don't have the necessary research base to start from.
>
> This is where so much of British industry falls down: it fails to do the background research so that when a new breakthrough in science comes up it is not in a position to exploit it.

Birchall's other principal interest which he is now developing is biomaterials. This is the study of how organisms make structural components (bone, shell etc.) from inorganic minerals. Another interest is bio-inorganic chemistry, which is the study of how inorganic elements, such as metals, interact with biological systems. It includes the study of how iron, silicon and aluminium behave in living systems such as animals and humans. The fairly recently publicised connection between aluminium and Alzheimer's disease has given the subject a much higher profile. So, too, has the discovery that the population in early Roman Britain suffered from lead poisoning due to their plumbing and central heating systems. But these are only small examples. Birchall believes that the subject will grow enormously in the next decade.

Two elements that particularly arouse his interest are silicon and aluminium. These are two of the commonest elements in the earth yet they have hardly received the attention that he thinks they deserve. Again, he asks the basic question: 'If aluminium is so toxic why hasn't the planet poisoned everything?' He suspects that it is because of the presence of silicon which can block the absorption of aluminium.

This theory, and Birchall's twenty-year fascination with aluminium and silicon, goes back to his fortuitous attendance in 1971 at a lecture on the importance of silicon given by a Californian professor, Klaus Schwartz. Schwartz showed that when silicon is taken out of the diet of animals they become ill. Birchall says that it now seems probable that silicon starvation encourages the aluminium in a body to become 'bioavailable', and therefore toxic.

Birchall's interest in silicon has taken him into the whole field of the 'trace elements', which are essential to our health. An example is selenium which slows down the oxidation of tissue. Too little of this and we age prematurely, he believes. This interest also overlaps in an academic way with some of Professor Roberts' work on enzymes. He explains: 'Many metals help us understand how enzymes work because most enzymes have metals in them and it is the metals that do the work. About a hundred and sixty enzymes depend on zinc.' Some of these elements, such as aluminium, also influence DNA.

Birchall says that his experience tells him that 'science has a funny way of creeping up on you.' A piece of apparently useless knowledge can suddenly become relevant when put alongside some new research elsewhere. Often it will lie untouched, gathering dust, until someone sees a connection between the discovery and some social and industrial requirement. An example he cites is the invention of polythene in the 1930s. Polythene was invented because some scientists wanted to find out if ethylene would polymerise under pressure, but having found that it did and that they could make long strings of it there appeared to be absolutely no use for it. No one saw it as a material that had any practical value until scientists working on developing the first radar systems recognised its potential as an insulator. Having made its debut in radar systems it was subsequently taken up and developed into an all-purpose, cheap material.

Birchall argues that 'relevance' emerges from scientific research and that it cannot be conjured up artificially. The basic research comes first and out of this, at a later point, comes an application. This mirrors what Lord Porter has said: that there are only two kinds of science, the applied and the yet-to-be-applied. Both these scientists are thus arguing that all science is useful in the end. One person who perhaps illustrates the truth of this is George Gray, formerly Professor of Chemistry at Hull University and now consultant research coordinator at Merck Ltd. For about fifteen years Gray pursued an interest in liquid crystals; for much of that time conducting his research virtually on his own and with very little funding because it was not regarded as being particularly important. In the latter part of the 1960s, his work came into its own when he was asked to design the material for what we now call a liquid crystal display. With the accumulated experience of hundreds of experiments behind him he was able to come up with a design pretty rapidly.

Gray began his research into liquid crystals as an Assistant Lecturer studying for his Ph.D. at Hull in 1953, and he was encouraged into this area by Professor Brynmor Jones. Jones had discovered that certain acids had the curious property, when heated, of showing crystal-like qualities even though they were in liquid form. This was a puzzle: the conventional wisdom was that chemical substances were either solid or liquid or gaseous. However, a hundred years previously it had been discovered that when the solid crystals of some substances were heated, rather than becoming immediately liquid they transformed first into an intermediate state or phase which had both liquid and crystalline characteristics and only later became a normal liquid. In this intermediate state the molecules were neither totally ordered – as in a crystal – nor were they disorganised as in a liquid. So there was some information about liquid crystals, but not much, and the relationship between molecular structure and liquid crystal behaviour had never been explored. Between 1953 and 1967 Gray studied this subject almost uninterruptedly.

His approach was to synthesise as many different chemical substances as possible and heat each one, noting at what temperature it changed from being a crystal to becoming a liquid crystal. In some cases they never formed a liquid crystal at all. By working in this step-by-step way, he says, 'one could make correlations between the molecular structure of materials and their ability to form liquid crystals of one sort or another.' He was thus building a catalogue of knowledge for the sheer fun of it. This work was done, he says, with very little financial support – 'At one stage I almost abandoned it.' Indeed for some of his time, he worked on anti-bacterial substances and germicides, and this research was financed by Reckitt & Colman, makers of household cleaning agents.

The breakthrough came in 1967 when he was contacted by Professor Cyril Hilsum of the government's Royal Signals and Radar Research Establishment (RSRE). The RSRE is a Ministry of Defence research centre, which has successfully been involved in research into radar, carbon fibres, gallium arsenide and many other inventions. Some US scientists at the Radio Corporation of America (RCA) had been working on liquid crystals and had realised that such materials might be used to present information in electro-optical display devices. If an electric field was applied across a thin film of liquid crystal of low enough resistivity, a turbulent, stirring motion was induced giving a different optical appearance to those parts of the film where the field was applied compared with those where it was not. Patterns, information, etc, could thus be displayed. It was a good idea, but there were practical problems associated with the lifetime of such devices.

A significant development was the invention by Schadt and Helgrich of the Twisted Nematic Liquid Crystal Display, requiring room temperature materials, of low conductivity which responded purely by a field effect, causing a realignment of the long axes of the liquid crystal molecules, and a consequent electro-optical effect. Gray's task was to produce stable organic liquid crystal materials that would perform in this way.

Gray says that when first faced with these problems he immediately had an idea how they might be resolved, because the basic science had been done, and molecular structure/liquid crystal property correlations had been established. Eventually a family of suitable liquid crystals was developed, which responded well to electric fields: they were stable and colourless and were not toxic.

Some of the pure materials were liquid crystals at ambient temperatures, but their temperature ranges were restricted. The inventory of available materials of related structures was therefore extended by Gray and his colleagues, and in collaboration with scientists at RSRE and BDH Ltd (now owned by Merck), new mixtures were formulated with properties eminently compatible with those required for the Twisted Nematic electro-optic display.

The new materials were liked by display manufacturers and the first liquid crystal displays were marketed in 1975. Today they are a universal, low energy consumption method of providing information in a variety of equipment ranging from watches, calculators and instrument displays to lap-top computer monitors and full colour television screens. The popularity of the displays has meant that there has been pressure to develop higher levels of sophistication, allowing more complex data to be displayed and the use of full colour. The original materials, and related families of materials, have ben developed in response to the market's demands and have secured a multi-million dollar industry in liquid crystal electro-optical displays.

A spin-off development was the design of a range of stable thermo-chromic liquid crystals, i.e. materials that change colour according to the temperature around them. This phenomenon has ancient origins, for the first such liquid crystal was discovered in 1888, quite by chance, by a German botanist, Reinitzer, who made observed it as a natural derivative of cholesterol. Thermo-chromic liquid crystals have become the basis for digital thermometers and other forms of thermal indicators.

Unsurprisingly, in the light of his own experience, Gray is another scientist who regards the division between basic and applied research as in some senses artificial. He derides the idea that academic scientists should not get involved in industrial problems. The interaction with customers is exciting and challenges you to devise new solutions. Yet at the same time he believes that the emphasis on applied research, especially in universities, has been taken too far. He recalls being asked by a German professor, while he was still at Hull University, how much of his research was applied and how much was fundamental. When he replied that 80% of his work was in applied research, the German professor replied: 'If I did that I'd be told that I wasn't fulfilling my academic duty to pursue new knowledge.'

Gray adds that the pressure on academics to bring in outside funds and do contract industrial research work has reached ridiculous proportions. He says, scathingly:

> It is probably easier now for a chemist to please his financial masters by running a chemical spectra on cooking fat for a local fish and chip shop and being paid £1,000 a month, rather than by getting funding for a solid three-year piece of peer-reviewed fundamental research, because such funding for basic research rarely covers the whole of the costs.

The pressure on academics to run after industrial funding is encouraging some of the best of them to leave the academic world altogether. Gray says that: 'Many academics are saying that if we are going to be under this kind of pressure to get into bed with industry we might as well join it and be properly paid.'

One development that has particularly excited chemists all round the world has been the discovery in the mid-1980s of a new form of carbon. Its molecular form is that of a ball structured like a pair of geodesic domes; this led Professor Harold Kroto of Sussex University to coined the name buckminsterfullerene, after the architect, Buckminster Fuller, who designed the first geodesic dome. This form of carbon is now called a fullerene, or colloquially a bucky ball. The commonest variety is carbon 60 or C60, made up of 60 carbon atoms.

The discovery of what has become a family of new carbons has astonished chemists for two reasons. Firstly, until the mid-1980s carbon as a giant molecule was only known in two relatively well-characterised forms – graphite and diamond. Secondly, the fullerene carbon has some very interesting properties: in its pure form it is immensely strong and could well provide the basis for new materials, but the structure of this carbon also suggests that, when combined with other molecules to form compounds it will provide a basis for new chemicals possessing great potential. So a whole new field of chemistry has opened up as a result of this accidental discovery.

Kroto has been one of the pioneers in this field. Like George Gray and others, he has had to struggle for financial support. His early research in this field was pursued against a background of cuts by SERC, and the lack of funds at a critical time during the research so delayed Kroto's progress that other scientists in Germany and the USA were able to advance some important aspects of this new field ahead of him. In this respect Kroto's experience is symptomatic of what is happening to British science.

Kroto says that his original interest was studying cyanapolyynes (unsaturated carbon molecules) in a laboratory, looking at their molecular dynamics. David Walton, a colleague of Kroto's at Sussex University, had developed elegant methods for synthesising long chain polyyne molecules. At about the same time (1974/1975) spectacular advances began to be made in an apparently unrelated field, molecular astronomy. Astronomers were studying the spectra of the cold dark clouds in interstellar space, and these appeared to contain some unusual molecules. This prompted Kroto and Walton, with Tcakeshi Oka and Canadian astronomers, to check whether the molecules that they had made in the laboratory existed in space. It turned out that this was the case, which deepened Kroto's interest in carbon molecules in space.

Carbon is one of the most common elements on our planet. Every living organism contains some carbon and this carbon in turn derives from the destruction of stars millions of years ago. So it was natural for a scientist like Kroto to be curious about the carbon out in space, and it prompted the question: what is the structure of the carbonaceous components in space?

The cyanopolyynes found in space had chains of five, seven or nine molecules. Kroto wanted to simulate the stellar conditions producing these carbon chains in a laboratory, (i.e. to simulate the conditions in a carbon star) and to see whether even longer chains might exist in space. Then in 1984, on a visit to Rice University in Houston, Texas he learnt that one of the scientists there, Richard Smalley, had developed a laser vaporisation apparatus which enabled scientists to blast atoms from the surface of a solid target. The machine allowed scientists to look at clusters of vaporised atoms using a mass spectrometer to measure the number of clusters and the number of atoms in each cluster. The scientists, Richard Smalley and Bob Curl, were using the machine to vaporise metals and other refractory materials, in particular silicon carbide. When Kroto visited his friend Curl he asked him and Smalley if they would like to collaborate with him in his research – into carbon chains – using their machine to vaporise graphite. Kroto believed that by using the machine he could simulate the chemistry of a carbon star and perhaps make the carbon chains that he and Walton and his astronomer colleagues had detected a few years earlier in space. Smalley and Curl were interested in Kroto's proposal, but they had no time to pursue it. It was not a high priority for them because it did not fit into the semiconductor work they were engaged on. (Their grants were for the purpose of looking into materials for semiconductors – carbon does not have semiconductor properties.) However, about eighteen months later they invited Kroto to the USA to carry out the experiments. Kroto was so excited that he decided to go to the USA there and then, arriving in the USA three days after receiving the invitation from Smalley and Curl.

Kroto carried out a series of experiments to vaporise graphite with the help of Jim Heath (a key experimentalist), Sean O'Brien and Yuan Liu, all students of Smalley and Curl. They saw the long chains of carbon atoms that they had hoped for. They also found something totally unexpected. This was, says Kroto, 'an extremely strong signal for a cluster with 60 atoms which seemed to indicate that a cluster of this number was very stable. There were also strong signals for clusters of 70 atoms.' (There were even other clusters from 30 to 100, which had been noticed by other researchers at the Exxon oil company and at the AT&T research centre, but they had not noticed the extraordinary nature of the behaviour of the cluster of 60 atoms.) These results were odd, to say the least, since each carbon atom usually bonds with three or four atoms. In diamonds carbon atoms bond with four atoms of hydrogen while in graphite they bond with three carbon atoms. In graphite and diamond the carbon structures can grow indefinitely in a flat lattice-like structures. But in these experiments the carbon atoms tended to settle in structures of 60 atoms or sometimes 70.

Kroto adds that to some extent Smalley's new apparatus 'was almost designed to discover C60'. This was because the instrument that they were using 'was designed to throw up clusters of a few tens of atoms which interacted with each other in an environment of inert gas.' Yet it was still a mystery how these clusters were being formed. Kroto and his colleagues eventually decided that they were forming into closed cages rather like a football. Kroto remembered Buckminster Fuller's geodesic dome at the International Exhibition at Montreal in 1967, and called the new carbon structures buckminsterfullerenes. Kroto, Heath, O'Brien, Curl and Smalley published a paper on their research in *Nature* in 1986.

Yet Kroto and his colleagues still had to prove the structure of this new carbon molecule, which could not be seen and which only existed during an experiment.

To do this they needed to make it in large quantities. Back in Britain in 1987, Kroto tried to do this with his student Ken McKay, and obtained some interesting preliminary results, but attempts to gain modest funds, around £20,000, to go to the next step were not successful. Kroto says he made numerous requests for funding to major chemical companies over the next two-and-a-half years. He adds that he had just been given £180,000 from the SERC to build apparatus similar to that at Rice University. This meant that he could hardly go back to the SERC and ask for more funding immediately, especially for something that seemed so speculative. For what he now wanted to do was 'an innovative departure, i.e. to look at arc processing'. He explains: 'McKay and I had used a carbon arc, and had seen a change in the structure of the carbon deposit, and I wanted to replicate this experiment and then use a mass spectrometer to analyse what was taking place prior to deposition.' So he had to give up progressing with this research. He adds: 'This is the sort of speculative research that doesn't get funded nowadays.'

Three years later two other researchers, Donald Huffman at the University of Arizona in Tucson and a student of Wolfgang Kraetschmer, Kostas Fostiropoulos, at the Max Planck Institute for Nuclear Physics at Heidelberg, who had been working in a related field since 1982, announced that they also had seen clusters of carbon with 60 atoms in their experiments. They had analysed carbon soot, which they had made by heating graphite rods electrically inside a bell jar partially filled with helium. They now speculated that what they had found were also the buckminsterfullerenes that Kroto, Smalley and Curl had discovered.

In creating quantities of carbon soot, Kraetschmer and Huffman had managed to do what Kroto had tried to do in 1987. Their experiment was almost identical to the one that Kroto had planned with Ken Mckay. The news of their success spurred Kroto to repeat his experiments with the help of a student at Sussex, Jonathan Hare, and that of Amit Sakar and Ala'a Abdul-Sada. He was supported in this by some extra money which Steven Wood of British Gas had allocated for research into combustion and C60. During 1989 and 1990 Kroto and his team in Sussex and Huffman and Kraetschmer were separately working on experiments not just to make the carbon soot but, more importantly, to extract the carbon 60 molecules from it. Kroto's team fell behind because all the time they were working with outdated, badly functioning equipment. Kroto adds: 'Under the current funding base it is not possible now for all departments to keep all their equipment up to scratch. Here, one result was that some of our equipment was frequently breaking down.' (The electronic circuitry of the arc equipment, for example, failed and was virtually rebuilt by Jonathan Hare.) Not surprisingly Huffman and Kraetschmer became the first team to extract macroscopic quantities of carbon 60 molecules, and they were able to measure its crystal structure. Kroto was stunned by this news, as the Sussex team had already identified C60 by mass spectrometry and, only days before, Jonathan Hare had succeeded in extracting a red solution which he and Kroto thought might be C60.

One more remaining piece of the jigsaw remained: proving the football structure of carbon 60 and its stablemate carbon 70 when formed from the rods of two kinds of carbon – carbon 13 and carbon 12. This study was an important refinement of the work of Huffman and Kraetschmer. With another Sussex colleague, Roger Taylor, Kroto's team further demonstrated the geodesic structure of the carbon 60 and carbon 70 molecules. They beat another team – at IBM – engaged on the same enquiry by seven days.

Fullerenes are now being made in laboratories all over the world, with potential users in the electronics and oil industry.

For Kroto the research has been 'an up and down experience'. There has never been enough money to pursue this research thoroughly, and he certainly feels that he might have achieved more, earlier, if he had not always had to worry about how to resource his group sufficiently for them to be able to try speculative ideas out rapidly. He says:

> Had we been funded on the third [arc processing] project in 1987, as had been vaguely promised by one company, there is no doubt in my mind that we would have extracted C60 within a year, and that company would have had every patent worth having on C60.

Commenting on this research in a paper written for a German chemistry journal in February 1992, Kroto remarked: 'This advance is an achievement of fundamental science, and serves as a timely reminder that fundamental science can achieve results of importance for strategic and applied areas.' He also pointed out that it owed a great deal to the inter-disciplinary nature of courses at Sussex University, (such as the Chemistry by Thesis degree course) where undergraduates have been able to 'to carry out research with supervisors from more than one field'. He added:

> Sadly, this and other courses have been 'regulated' out of existence by bureaucrats who have little understanding of how student research expertise is brought to maturity, and no awareness of the dire consequences for our future scientific capability.

These four examples show some of the achievements of basic science, but they also highlight some of the problems in funding it. Funding from government sources has been severely squeezed, and we cannot look to industry to make up the difference. Industry in the present economic climate is becoming more reluctant to invest in fundamental science whether it is internal or external. Even in the highly research-driven chemical industry, both ICI and BP are cutting back on research, much to the consternation of people like Professor Birchall. Meanwhile many scientists for their part are being forced to compromise their research objectives for the sake of short-term industrial goals.

Industry will certainly fund important work where an area of science is seen to be immediately relevant, as in the examples cited by Professor Roberts. But if a subject lies outside industry's immediate peoccupations then it gets pushed to one side. There lies the problem with the Thatcher-driven argument for targeted, sponsored science. No one wanted to fund liquid crystal research in the 1950s and early 1960s, because there was no obvious use for such research. Gray remembers being told in 1967 that 'liquid crystals may have a minor role for displays in high ambient lighting, but they will make no impact on black and white or colour television.' But just look what is on your television control system or your video recorder, or your lap-top computer screen: it is a liquid crystal display. Innovation in chemistry rests just as much, if not more, upon basic research as it does on the application and modification of existing knowledge.

The Kroto story is illuminating too because its features are replicated everywhere in the current university research system: first-class scientists often working with out-dated equipment, suffering delays in reaching their results and therefore falling behind their competitors. In these circumstances British science is diminished, and so too is British industry which is one of the main beneficiaries of science. Our chemical industry is one of the few sectors that still ranks among the world leaders, and which brings in £3bn surplus a year to our balance of payments. But, as everyone in the industry points out, it relies on high-quality university research to maintain its position and to maintain the jobs of its employees.

8

COGNITIVE SCIENTISTS: OF MINDS AND MACHINES

We know that the brain is made of the same stuff as the rest of nature, and its atoms must therefore obey the same natural laws as other atoms do. In that sense, then, it is tempting and even reasonable to say that the brain must be some kind of machine. But to use the word machine . . . misses the crux of the question. The real question about the human mind lies deeper: is the brain a machine with a formal procedure of any kind that we can now conceive?

Dr Jacob Bronowski, addressing the American
Association for the Advancement of Science in 1965 on
'The Logic of Mind'.

The kind of question that Dr Bronowski poses above has long intrigued 'hard scientists' as well as philosophers and psychologists. In the last forty years it has led to two fields of research. One is the investigation of the human and the animal brain from a neurophysiological and psychological point of view while the second is the study of the computer brain from the perspective of computer theorists and designers.

These two fields of study have much in common, being concerned with concepts such as perception, memory, the organisation of information, logic and communication, and they have stimulated each other. Indeed they are increasingly seen as complementary. Both fields have implications for philosophy and psychology, while neurophysiology also overlaps with chemistry, biology and medicine. In the last fifteen years studies of human brains and of the branch of computer science known as 'artificial intelligence' have made considerable advances. In neuroscience, for example, we now know which parts of the brain contain the memory and which parts are involved in perception, and it is now recognised that the two halves of the brain influence quite different aspects of behaviour and thinking.

Meanwhile computer scientists have designed some computer programs that are able to reproduce some limited aspects of human intelligence. In pushing forward this research we have learnt more about the human mind and expanded our understanding of the potential of the computer. The research that lies behind these discoveries is therefore not just an idle academic pursuit but holds out the prospect of significant applications in medicine, computing and even in agriculture.

This research is funded by a number of bodies such as the Medical Research Council, the Science and Engineering Research Council and several charities. This reflects the fact that the research spreads across a number of different disciplines, as two of the examples in this chapter show. The study of human and animal brains, for example, probably embraces more scientific disciplines than any other field of science.

At the core of this multi-faceted research lie a host of formidable questions. These include: What is mind? What is consciousness? What is the relationship between the mind and brain? To what extent is the brain like a computer and can a computer mimic the processes of the brain? How do perception and memory work? How do chemicals in the brain facilitate or inhibit its functions and how does the brain renew itself?

One person who has long examined these questions from a vvariety of angles is Richard Gregory, Professor of Neuropsychology at Bristol University. Like most scientists he is fond of puzzles, in his case the puzzle of perception. His inter-5est in how the brain works has inspired him to write a number of books: *EEye and Brain*, *The Intelligent Eye*, *Mind in Science*, and *The Oxford Companion to the Mind*. But his views are not just theories, they have been shaped by experiments and some unusual observations.

An early observation was watching a blind man aged 52 come to terms with the renewal of his sight after an operation. He had been blind since the age of ten months, perhaps earlier. What was interesting about this man was that he could see things immediately that he already knew by touch, but he could not see things that he had not experienced in this way. For example, he could recognise capital letters, but not lower-case ones, and it turned out that at blind school the chidren had been given raised letters on wooden boards, over which they could run their fingers, but that these had all been capitals. Gregory was amazed. What he saw led him to a number of conclusions:

> It meant that knowledge is necessary for seeing things. It meant that there is a common knowledge base that the senses can draw upon and that this knowledge is crucially needed in order to see. It also means that seeing and knowing and seeing and believing are very closely related.

This revelation led Gregory to ask further questions which he is still asking today. One is that if knowledge is required for seeing, how does the brain deal with illusions, how do perceptual errors occur? Gregory now says that he regards 'each perception as a hypothesis set up by the brain'. The brain starts with a hypothesis based on bits and pieces of information derived from sensory information coming in and on information already stored, and then it tests the hypothesis which it either accepts or rejects; if it rejects the hypothesis then the it seeks to arrive at a better one. This process, says Gregory, is analogous to the scientific method itself with its dedication to enquiry, experimentation, analysis and the synthesis of information.

This view of the brain leads to an obvious question. If the brain works with stored information does that mean that it therefore works like a computer? For a long time Gregory says that he was seduced into thinking that this was so. He became involved in some of the early work in Edinburgh on artificial intelligence, i.e. the design of computer systems that can simulate human intelligence. Gregory now believes that to see the brain as a sort of computer is mistaken:

> It's the wrong way of looking at, because the brain doesn't go through specific programs like a computer. A computer works in a step by step process, whereas the brain settles into stable states representing answers from repeated experiences of situations. But the brain is certainly a physical system.

One of the difficulties in trying to work out how the brain functions is in deciding what the brain is. How can you begin to ascertain what the parts signify and what they do, if you are not sure what the overall system is? How can you perform experiments on the brain if you are unsure what you are experimenting on? This has been another puzzle for Gregory. If one accepts that the brain is a physical system, could one therefore do experiments on it as if it were an engineering system? He explains:

> I wanted to compare the brain with a machine and to find out whether they were logically similar or different. Could you compare the brain with a radio which is made up of a number of component parts? Could you specify and isolate important functions.

His answer was yes and no. He says:

> The problem in trying to specify functions by removing them is that you disturb the system and you can reach the wrong conclusions. If you remove a component from a radio and it whistles that doesn't mean that the component's function had been to stop whistling. So you can't deduce a function by removing it and seeing what happens. The same is true of the brain.

One piece of technology that has come to the rescue of this kind of research is a specialised brain scanner called a PET scanner (Positron Electron Tomography scanner). This has enabled scientists to study the biochemistry of some of the modules in the brain that govern perception or memory. These new scanners are 'absolutely marvellous', says Gregory. 'You can now see which parts of the brain light up according to the function being used, and these observations have been corroborated by stimulating bits of the brain electrically.' But what we do not yet understand sufficiently are the underlying processes – the molecular processes – involved in the various parts of the brain; nor do we fully understand how different parts of the brain interact with each other.

Gregory's interest in perception, not to mention philosophy, psychology, history and much else, has shaped his views on the public perception of science in Britain. In some ways, he is not surprised at the lack of understanding. Science is difficult, he says, and it often requires at least a passing familiarity with mathematics with which we seem to be uncomfortable. To understand science, people need 'first-hand experience of it', he says, which is why he set up a science centre in Bristol called the Exploratory where the public can experience some aspects of science for themselves. Gregory, who was taught by Bertrand Russell at Cambridge, remembers Russell saying that there are two kinds of knowledge: knowledge by acquaintance and knowledge by description. To appreciate science, you really need both.

In order to gain a better understanding of its general character, some scientists are now focusing on specific functions in the brain, and they are looking both at the mechanics of what is going on and at the molecular basis of the brain's activity. One such person is Colin Blakemore, Waynflete Professor of Physiology and Director of the MacDonnell Pew Centre for Cognitive Neuroscience at Oxford University. In the mid-1980s Blakemore presented a series of television programmes under the title 'The Mind Machine', in which he probed our understanding of the brain from both a modern and a historical perspective. Blakemore, who is a former student of Gregory's, has long been fascinated by the process of vision in humans and animals. The visual system, he says, is an excellent model for trying to understand many general properties of the brain 'simply because we know so much about it', adding that it has given us clues to the organisation of the cerebral cortex and to the principles underlying the development of the brain.

His interest has been at the neurological level, and it has been directed at solving questions such as how we form three-dimensional perceptions and how we learn to interpret objects, especially early in life. From there his thoughts have turned to notions of consciousness and individuality. But the subject widens out even further to embrace physics and genetics and has found applications in robotics, computers and educational development. His initial interest, he says,

> was to explore the neural basis of perception: what links the activity of nerve cells with the way we actually see things. It's a philosphical, indeed a metaphysical question, and it touches on the whole concept of consciousness. I felt that the properties of nerve cells might give insights into psychological phenomena, such as visual illusions and even more mysterious processes like depth perception, which is the recognition of three-dimensional shapes. The psychologists had already shown that the brain can put together information from different sources through subtle interpretations of the retinal image. I wanted to understand this process in neurophysiological terms.

It was the three-dimensional quality of vision that particularly intrigued Blakemore. The brain has the task of interpreting a flat two-dimensional image in each retina in terms of the objects out in space that have given rise to those images. That process, says Blakemore, 'is far more complicated than we can imagine. It's not like taking a photograph, because it has to be interpreted.' How this is achieved became his great interest.

An early piece of research, done at the University of California at Berkeley, looked at the basic process of distance perception called stereoscopic vision, i.e. how information from the two eyes is combined. Blakemore worked with two other scientists, Jack Pettigrew, an Australian studying medicine at the time, and Horace Barlow, a great-grandson of Darwin. The research examined the way in which the brain is able 'to compare minute differences in the geometry of two retinal images', and succeeded in finding the mechanism in the visual cortex. The scientists actually found the very nerve cells that 'could measure the distances of objects in space'.

Subsequent research in Cambridge with another eminent scientist also working in this area, Fergus Campbell, showed how the brain might analyse the distribution of light in the retinal image. The conclusion of this research was extraordinary and influential. As Blakemore explains: 'We showed that the visual image decomposes the scene in an analytical way and in a way that can be understood by physicists to derive the local information about light distribution.'

The results from this research, and of other similar work, extend beyond the interpretation of human vision. They are starting to play a part in the design of robots. As Blakemore explains:

> If you want to make an artificial eye for a robot it is not enough to have a camera and simply create a two-dimensional image of the world. You have to have a mechanism for interpreting that information and reflecting on its meaning.

That mechanism will be derived from studying the human brain.

As an ex-student of Gregory, Blakemore has always been well aware of the story of the blind man who regained his sight. For him the story was fascinating not only for what it said about the process of perception but also for the questions that it posed about the development of perception. He says:

> The machinery of vision as revealed by perceptual and neurophysiological experiments is so amazingly elegant that I wanted to know whether it was innately made that way, constructed entirely through genetics, or whether experience could program the brain's mechanism for responding to the outside world.

Gregory's blind man had been able to see as an infant, but how had that child seen the world? During the late 1960s Blakemore began to wonder what powers of sight that young child would have had. How had it seen the world as a baby, how had its vision as a baby been 'influenced by what it was seeing'?

To answer these questions, Blakemore reared two kittens in 1969 in environments in which each one could see only horizontal or vertical lines. He knew already that in cats and monkeys particular nerve cells in the visual parts of the cerebral hemispheres respond to lines at particular angles. The brain decomposes the visual scene into the constituent angles of the shapes in view. The point of the experiment was to find out if this facility was learnt or if it was innate. So the kittens were reared in the dark and when put in the light they were exposed either to horizontal or vertical lines. Later on, when recordings were made from nerve cells of the visual cortex, 'there was a distinct lack of cells responding to the lines of angles that they had not seen when young.' This suggested that the visual capabilities of kittens 'were at least partly programmed by early experience'. Blakemore published his results in the journal *Nature*, arousing considerable scientific controversy; the ensuing debate lasted for more than five years. There was disagreement on the interpretation of the results, with some scientists suggesting that maybe the visual cortex of these kittens had originally been perfectly normal and that their inability to see certain kinds of lines was due to atrophy, and not because they had never been exposed to a particular kind of line. The evidence now, says Blakemore, supports his original contention that perceptual capacities are partly learnt. This also confirms Gregory's conclusion that seeing depends on knowledge.

The more important aspect of this research is that it has a message for our understanding of the rest of the brain. It provides another example to show that 'nerve cells can change and reorganise their connections'. This plasticity of nerve cells is one of the subjects that now preoccupies Blakemore, and it is one of the areas of research being undertaken by some of the teams working in the McDonnell Pew Centre at Oxford. The Centre, which was set up in 1990, is a multi-disciplinary one and brings together Oxford academics from the fields of philosophy, cognitive psychology and neuroscience with physicists and engineers interested in fields such as robotics. In simple terms it will concentrate on what is now called cognitive neuroscience, i.e. the study of how the brain acquires and organises its understanding of the world and how this function is developed or impaired by brain damage and disease.

Cognitive neuroscience is becoming a dominant theme within brain research, and it clearly fascinates Blakemore. He explains:

> The brain is the only organ able to escape from the tight constraints of genetic determination; it is able to modify itself on a scale of complexity that isn't comparable with any other organ. It is now known that certain parts of the brain are capable of plastic change, where there are mechanisms in the nerve cells for rapidly changing the strengths of particular contacts from other nerve fibres on to those cells. The cells can regulate by simple algorithms of interaction with their inputs which of those inputs will be strengthened or weakened, depending on how active each connection has been. Thus the brain can change its own circuitry. Now the major issue concerns the molecular mechanisms that control this process.

This research touches on the old nature versus nurture debate and in the view of Blakemore it even has implications for educational policy. As he explains:

> A fundamental question and concern is the extent to which the differences between

people reflect differences in their genetic make-up and to what extent these differences reflect the various ways those brains have been used.

If we are entirely what our genes make us, then education is pointless. Therefore knowing how that learning happens is of great importance, that is, especially, knowing the critical times at which the brain is capable of learning particular things. If we could define for each elementary intellectual function what were the optimal timings and conditions for changing the appropriate part of the brain, then we could design an educational system that is firmly rooted in biology and in our understanding of how brains really work. If you look at those areas of the brain that we do understand, that knowledge doesn't justify very well our educational system.

As an example he cites language acquisition which occupies a sensitive period in our lives that ends at about the age of nine. The inference is that children should spend their early school years learning languages, delaying the teaching of other subjects for which the 'sensitive period' of learning occurs at a more advanced age.

As with much scientific research new knowledge in one area of science can often be relevant for understanding another area. The field of neuroscience is an interesting example. In talking about the plasticity of the brain Blakemore conjectures that

It is likely that the adaptive mechanisms in one part of the brain are the same as in others – they've been developed through evolution. We see more and more the continuity of fundamental molecular mechanisms throughout the body and throughout the living world. Once evolution discovers a handy trick which depends on a particular molecular mechanism it will exploit it again and again.

So the adaptive molecular mechanisms in the visual system being researched by Blakemore and his team in the UK and by researchers in the USA may well reveal information that is important for other parts of the brain and just possibly for other parts of the body.

Meanwhile at Sussex University another line of research is being pursued. Like the McDonnell Pew Centre at Oxford, the Sussex Centre for Neuroscience is multi-disciplinary. The Centre is one of the newest 'inter-disciplinary research centres' (IRCs) set up by the Science and Engineering Research Council, which is putting £4m into the research over a four-year period. Like all the IRCs the Centre's terms of reference stipulate that it must draw in some additional funds from industrial clients. So the work here has two aims: to provide a source of fundamental knowledge and to provide research that can be applied within a relatively short time-scale, i.e. over the next year or two. The Director of the Centre is Professor Michael O'Shea.

Unlike the other scientists mentioned so far O'Shea is a biologist with a particular interest in molecular biology. He also has a deeply personal interest in researching the brain because his daughter died of a brain tumour in 1990 at the age of eleven.

The core of this research will look at nervous systems, not in humans, but in very small invertebrates such as molluscs and insects. This may seem an odd starting point but there are good practical and historical reasons for beginning research on small animals, as O'Shea explains:

> In a vertebrate or a human brain the population of cells is too large to analyse the whole neural system in terms of the interaction between the component parts. So the answer is to reduce the problem to a system that has a few neurons [nerve cells] that interact with each other so that you can understand something of the fundamental rules that govern nervous systems in general.
>
> At the molecular level there is no difference between the nervous system of a simple organism and one that is more complex. What is different is that more advanced organisms have a larger number of neural components. When you look at simple organisms and see the chemical messengers involved, what cells are involved in a particular behaviour and what are the electrical and chemical processes involved in information transmitting systems, there is no great difference between how they do it and how we do it. Sometimes they use identical chemicals to us.

This is another case where research in one area is highly applicable in another.

The use of relatively simple animals to study the nervous system goes back decades. A number of important fundamental discoveries were made on invertebrates. For example the discovery of what biologists call the 'action potential' (the electrical pulse which by which one nerve cell communicates with others) came through the study of squids in the 1940s. The basic principles that generate the nerve impulses in a squid apply to almost all living creatures. O'Shea adds that, apart from the squid, much of what we know about the brain's processes were first analysed in organisms such as the crayfish, the snail and the locust.

What is interesting about the project at Sussex is that it brings together scientists from the disciplines of chemistry and biology. This is important, says O'Shea, because until recently the emphasis in understanding nervous systems was on the electrical properties of the neurons with less emphasis on their chemistry. By bringing both these disciplines to bear on the subject the aim is to integrate the two approaches. A key element of the research will be to understand the chemical mechanisms employed in various parts of the nervous system: how the receptors through which cells receive chemical information work; what chemical interactions are involved; what determines how the neurons interconnect with each other and what genes are responsible for these functions. With ambitions such as those it is not surprising that the research is intended to last at least four years.

O'Shea describes the research approach as 'a kind of cellular and molecular dissection', aimed at contructing a huge chemical and electrical wiring diagram in which one can analyse each component's function independently. He agrees with

Gregory that with a large vertebrate or human nervous system this kind of analysis would not be feasible, but with very small nervous systems the problems become more manageable. What the diagram of a researched nervous system will look like after one year's work is impossible to say. Even forecasting how the research will progress is also difficult to assess, which means that 'planning' a project in this context takes on a different meaning from what it would mean in other circumstances. O'Shea explains:

> You write a research plan, but it is inevitable that you will go off on tangents. Part of the skill is detecting where to go in an ongoing project where the goal you set initially may appear less interesting than when you set it. With experience you learn to judge when to go off on a new line.

Planning a long piece of research is therefore not simply a question of targeting a goal and sticking to it, partly because in some circumstances it may be foolish to do so:

> Sometimes you can set goals for yourself that become not just difficult but impossible so you'd be wasting your time. A judgement often has to be made about how much tenacity is really good for you. There comes a point when you can try too hard, when you have to change direction. There's a lot of change of direction that happens in basic research which people who are not in science don't appreciate.

There is a paradox in neuroscience research, says O'Shea, which is that the machinery required to conduct this research is the brain itself. 'There might be a problem here, for it might be impossible for the brain ultimately to explain how itself works.' So although this research programme touches on some of the most basic questions that we can ask about ourselves, such as what is consciousness, it is arguable that it may never be able to answer them. The brain in effect may be unable to provide a satisfactory explanation of itself. Yet if the research can explain some of the processes that lie below consciousness and control behaviour and moods, which influence sensory perception and which integrate the information that comes into the neural system, then it will have answered some very important questions.

Apart from the intrinsic scientific interest of this research, there are also a number of possible applications. One is that it may enable us to understand more about diseases of the central nervous system such as Parkinson's disease and motor neuron disease. This knowledge might then be used in the design of new drugs or even genetic treatments. Another application, more realisable in the immediate future, is that it may enable us to design new pesticides. These pesticides would be directed both against crop pests and against disease-carrying insects. They would target specific parts of the insects' neural system in a way that has not been possible up till now, and they would be so specific that they would not harm other species. They might work by mimicking or blocking a chemical such as an enzyme which acts on small protein molecules to produce peptides. The peptides are involved in a range of activities from metabolism to reproduction, and each one has a specific function so it would be possible to interfere with the synthesis of one particular peptide in order to destroy an insect.

So, alongside their basic research, the team of Sussex scientists will also be carrying out contractual research work for industrial clients, initially agrochemical companies. The nature of this work will probably be both research and development i.e. testing out certain ideas and also designing solutions.

A colleague of O'Shea's, Richard Andrew, Professor of Biology, is studying memory patterns in small animals, mainly birds. There are some parallels between his work and that of Blakemore: both of them are concerned with vision, memory and thinking. Andrew is immensely curious about the way in which the two halves of the forebrain of small animals work together. He adds that, since the brains of animals are not so very different from our own, gaining an understanding of an animal brain can enlarge our knowledge of the human brain.

Like Blakemore, Andrew is interested in memory formation. He explains:

> A bird will use one side of its brain to make an assessment, for example, about a piece of food, and it will use the other side to make decisions about what it has observed. The question is, how are those two bits of information put together?

Andrew believes that there is a strong timing element in this process, during which information is compared, reconciled, changed or maybe just cross-filed. He has not yet by any means completed his research, but he thinks that these timing mechanisms are different in each of the two parts of the brain. It would explain why many animals appear to remember, then forget and then again recall information. A bee, for example, will learn something about a flower and then apparently forget it seven minutes later, only to remember it again some time afterwards. The process by which animals and insects such as birds, rats and bees temporarily forget information has been 'completely mysterious', he says. What has excited Andrew is that he has found evidence to show that memory storage and retrieval in the two halves of the brain operate according to different timing systems. The fascinating question is how are the memories in the two halves of the brain brought together?

The workings of neurophysiological systems, as revealed by scientists, are infinitely more elaborate and flexible than any existing computer system, but some scientists believe that by studying them we may gain insights that will help in the design of computers and robots in the future. Professor Andrew: 'Biologists are asking if we can transfer some of the knowledge that we have about flies, for example their ability to stabilise themselves in air, to the design of robots.' Blakemore makes a similar point in relation to the study of the human brain:

> Human beings can do things undreamt of by computers. We have much better memory access, greater capacity to solve intellectual problems, greater versatility, greater physical compactness and our brains are able to repair themselves. Imagine if

we could discover the principles of those things, especially the way the brain programmes itself, then the pay-off for computers would be immense. I don't think this is science fiction. We will be designing computers based on brains very soon. On a small scale it is happening already.

One scientist who has been doing just that is Igor Aleksander, Professor of Electrical Engineering at Imperial College, London. In 1981 he and a group of colleagues at Brunel University designed a 'neural network', a computer modelled on our knowledge on our knowledge of biological networks of neurons – or brains. The machine, called Wisard, scanned images from a television camera and 'learnt' to recognise them. It was then able to compare them with unknown images and to recognise when an image was repeated or deformed. These images could be of people or objects. So the applications were considerable and the Wisard machine has been quite a commercial success

The difference between conventional computers and neural networks like Wisard is that conventional computers operate by programming whereas Wisard, in effect, constructs its own program by learning from its experience: 'the machine is given patterns and it works out its own rules.' It is no coincidence that computer scientists like Aleksander and physiologists like Blakemore and O'Shea both use the word neural. Both are essentially talking about brains, though admittedly of a different order.

This whole field of intelligent – or what Aleksander calls cognitive – computers has interested scientists since at least the time of Alan Turing, who worked on the idea in the 1950s and stimulated a whole generation of scientists in the USA and Europe. To people like Aleksander this is a rich area for research and speculation. As he explains: 'I was intrigued by the idea of learning versus programming in computers. There are so many things for which you can't write simple rules.' He describes Wisard as a device 'which extends what a computer can do by giving it an ability to learn from examples, what we call case-based reasoning. The Wisard machine reasons from a series of cases.' To write a program that would enable a conventional computer to do what Wisard does would take a programmer about two months, says Aleksander. But Wisard can be programmed in minutes, simply by 'showing' it a set of images, and can be reprogrammed for a new task equally easily. The machine is being used for a variety of applications including quality control on industrial production lines (the machine picks out defective products or components), safety and maintenance in power stations (the machine reads the film output from video cameras and compares what it sees against what it is supposed to see), quality control in computer component manufacture and Home Office security systems.

The inspiration for neural networks is quite simply the human brain. As Aleksander explains:

> We are trying to use the properties of bits of the brain to engineer our own systems which then end up doing things similar to the brain. We get cues from the brain and

then optimise the information with engineering.

But he qualifies this by adding that such machines can capture only an infinitesimal part of the brain's skills:

> What the brain and neural networks have in common is that they are both founded on a common theoretical base. If we didn't know anything of how the brain works it is most unlikely that we could come up with the structures in neural networks. They are based on informational laws from the brain which have a mathematical base.

One role which Aleksander foresees for neural networks is in helping to make some of our technology-based machines more user friendly. Many of the products that we use daily have more technology in them than we can understand. Examples are the video recorder, the personal computer and the switching devices on office telephones. We tie ourselves up in knots trying to make these things work. If we could design neural networks that understood our own language and if we could fit them as accessories on to some of our home and office gadgets, then just think how much more easily we could use them, says Aleksander. Even without the aid of a linguistic recognition capacity Aleksander believes that neural networks could help us to adapt some of our machines to our own particular preferences. But, although many scientists all over the world are working on these problems and have been for a decade or more, Aleksander reckons that it will be seven to ten years at least before neural networks make a significant impact.

Even so, some companies are already planning to use neural networks in their products. Aleksander says that the Japanese want to use neural networks in their consumer products. Some of the funding for his research comes from Japanese companies. This complements the funds that he receives from some other industrial sources as well as from the SERC.

Although Aleksander and his colleagues are surviving rather better than those in many other university engineering departments (his department ranks among the top ten in Britain), he fears for the future of his subject in this country. He says:

> In the last five years there has been a dearth of funding either for fundamental or applied research. The quantity of research is less and so is the quality. The quality has declined in terms of discovering new principles: there is a lot less fundamental research going on than there could be in terms of our potential. You see a lot of really good people not being funded. In the last three to four years only about 10% of the top class alpha-rated projects have been funded, which is quite ludicrous.

He adds: 'Politicians say that they don't want to fund high-risk science, but if you say that it takes you out of funding science straight away.'

In neuroscience there are compelling reasons for continuing our research efforts, especially at the molecular/genetic level, but they are rarely spelt out. Blakemore passionately believes that we need to support this research in the interests both of medicine and the economy:

> Nearly all the major brain diseases such as Alzheimer's, depressive disorders,

schizophrenia, Parkinson's and motor neuron disease have their basis in molecular biology and in genetics. The key to understanding them is in the genes. Those are the medical reasons for neuroscience and molecular biology, but there are others that are also important. The relationship between the drug manufacturers and the molecular biologists has become extremely close. This is why countries like the USA and Japan are piling money into these research areas, because there is the possibility of real root treatments of the basic brain disorders. So research has assumed an immense importance commercially. We seem to have forgotten this fundamental fact of commercial life that tomorrow's products are based on today's research. If today's research goes, then tomorrow's products go too. In many areas of neuroscience I think we have already lost out. We can't afford to lose any more.

9

PHYSICISTS:
THE SEARCH FOR THE SMALLEST

*When I try to teach special relativity, I am frequently
surprised by how little the revolution of Galileo or Newton
has entered students' minds. After years of teaching at
school, and possibly even after teaching at university, the
relativity of velocity and the absolute nature of acceleration
are still somewhat alien to most people's way of thinking.*
Professor Sir Herman Bondi, physicist, mathematician
and former Master of Churchill College, Cambridge.
(From an essay in *Let Newton Be*, edited by Fauvel,
Flood, Shortland and Wilson.)

If the some of the theories of Galileo and Newton are difficult to understand, the
theories of their successors such as Einstein often seem impenetrable to the lay
person, and even the experts are still grappling with the implications of early
twentieth-century physics. But it is undeniable that the the work of physicists
over the past hundred years has yielded a rich harvest for us all. The jet engine,
radar, the micro-chip, the fax machine and electronic touch controls in the kitchen
are just a few random examples of the benefits we have derived from work which,
when it was first published, appeared to be of no more than academic interest.

The concerns of physics are enormously broad. On the one hand, there are the
theorists, the nuclear physicists investigating the fundamental forces of nature
and the cosmologists attempting to understand the very structure of the universe.
On the other, there are also physics researchers whose work may well find its way
into products that will be on the market within a year or two. But although a great
gulf might seem to separate these two extremes, history has shown that, without
the work of the former, who have often appeared to be living in the tallest of ivory
towers, it would not have been possible for the latter to produce the machines
upon which we all rely.

British physics research is supported by the SERC. The largest single chunk of
funding comes from the SERC's nuclear physics budget which amounted to £85m
in 1991/1992. This may seem quite a large sum, but in fact £58m of the total was
used to pay subscriptions to international research facilities, such as CERN in

Geneva. The amount of money left for research grants and postgraduate awards was not much more than £10m; which, when spread across the entire university system, left many academics without any money to pursue fundamental research in nuclear physics. In addition some funding comes from the SERC's science board budget – around £15m to pay subscriptions to two other international research facilities and more than £5m in the form of research grants. Even with these additions the total physics research budget is considered by scientists to be too small.

The difficulty of obtaining funding, especially during the last five years, has led many physicists to leave Britain, among them Paul Davies who describes himself as a mathematical physicist. Davies left Newcastle University in 1990 to take up a post at Adelaide University. He is the author of several books such as *God and the New Physics* and *The Mind of God*, and, from an early age he was interested, he says, in questions such as the nature of consciousness, the nature of time and the origin of the universe. He did not see religion as a way of getting answers, so instead he turned to physics. He says: 'What has always attracted me to physics is the fact is that it is likely to provide answers to some of the deep questions of existence.'

Like many physicists, Davies has been deeply curious about how Einstein's relativity theory, which is essentially about gravitation, can be combined with the laws of quantum mechanics which are about how matter behaves at the sub-atomic level. Each of these theories, on its own, is mathematically consistent and allows physicists to make accurate predictions. However, all attempts to combine them and construct a larger theory that would account for the behaviour of matter on all scales, from galaxies to the particles within the atomic nucleus, have thrown up 'contradictions and mathematical inconsistencies'. The great challenge for physicists such as Davies is to try and reconcile them.

One route Davies has taken has been to examine 'cases where space-time structures, that is, gravitational fields, have created quantum effects'. One such instance is that of a star that collapses to form a black hole. Due to quantum-mechanical processes, as shown by Stephen Hawking, the black hole, from whose gravitational field nothing can in theory escape, will in fact produce radiation. Davies has been working on this phenomenon, and a related one by which, it is calculated, particles were created in the Big Bang by changing gravitational fields. In essence therefore, Davies describes his interest as investigating any example where 'gravity and quantum particles tangle with each other', researching it with mathematics and then exploring the consequences. Reflecting on those consequences has led him to write his books.

Davies' view of our universe is one that combines determinism and unpredictability. The universe, he believes, is neither completely mechanistic nor is it totally random. Quoting the French geneticist, Jacques Monod, he describes the universe as a compromise between necessity and chance. Our universe, he says, is ordered like a chess game where there are laws but where at the same time the outcomes are left open. There are some physicists, such as Peter Atkins of Oxford University, who disagree with him; they say that our universe is the way it is because it could not have been otherwise. By contrast, Davies says that 'mathematical physics shows that models of other universes are possible.' He adds:

> Given that there is an infinite variety of possible universes and that this universe looks to be so special in so many of its features, this indicates to me that a selection has been made.
>
> Either we accept it as a brute fact in which case the universe is a total mystery, or we assume that there is a reason why the universe is as it is, and that reason gives our lives and our particular universe a purpose.

In formulating his ideas, Davies has been influenced by a number of new theories. These include the notion of an inflationary universe, i.e. that the universe emerged from nothing as a quantum event and then jumped rapidly in size; the theories of Hawking and Hartle which suggest that it is possible for a universe not to be of infinite duration and yet for there to be no first moment of time; and also theories of chaos and theories of self-organising systems. All these developments have led Davies towards affirming an evolving, self-determining universe which is nevertheless constrained by the probability theory embedded in quantum mechanics.

Davies has also worked with other scientists such as Adrian Ottewill at Oxford University's Mathematical Institute. Professor Roger Penrose, who also works in the Institute, is investigating this field as well – trying to bring together the two main physical theories of this century, quantum theory and relativity theory. Penrose, who earlier in his career collaborated with Hawking, believes that 'there will be a big shift in our view of what reality is all about,' once we are able to unite these two theories. Already, he says, quantum theory has radically shifted our view of reality:

> One has to accept that at the level of quantum phenomena, alternative possibilities somehow co-exist, and that's a funny view of what is going on, but that's what quantum theory tells us.

The great puzzle for Penrose is the existence of 'time asymmetry' in physics, i.e. there is no distinction between the past and the future in the equations of quantum mechanics which apparently would apply equally well to a universe in which time is running backwards. Yet elsewhere in physics the second law of thermodynamics, which relates to large systems, tells us that systems become more disordered over time. How can those two theories co-exist, asks Penrose. 'How can you have something that is so blatantly asymmetrical in time, which seems to be built out of ingredients which are symmetrical in time?' Penrose's approach is to work back in time and find out what it was about the initial state of the universe that caused this apparent paradox to come about.

Neither of these physicists is an active experimentalist, but their theories will in the end stand or fall when put to the test by the particle physicists who are using massive accelerators to probe the fundamental structure of matter. In their giant machines they are striving to re-create, as nearly as possible, the conditions that existed in the very first moments of the universe's existence and thus, they hope, to discover some basic, fundamentally simple laws underlying the apparent complexity of the sub-nuclear world. Already, it has been shown that particles such as the proton and the neutron, once thought to be fundamental and indivisible, can be thought of as being made up of even smaller particles and, if we are searching for unity amid diversity, says Davies, then it will be among small particles such as quarks and gluons that the unity will be discovered.

One physicist experimenting with particles is Roger Cashmore, Professor of Physics at Oxford university. As a particle physicist his interest is the structure and origin of matter. He says:

> There are so many unanswered questions in the physical world and I like being part of the process that is attempting to answer some of them. For example, where does matter come from? Why is there no evidence that there is anti-matter in the universe? If the universe started with energy, then to create matter you have to create anti-matter as well but there is no sign of it. There's no sign of it experimentally. We see protons – which we're all made of – but not anti-protons. That is of enormous interest to physicists and cosmologists and indeed to astronomers. That is something that has really caught my imagination. . . I don't just want to take the world at face value and just sit back and say that's the way things happen. One wants to know why.

Cashmore is interested in the contradictions and puzzles of physics. He says: 'When you look inside the nucleus of an atom you see a proton and you find that it looks similar to a neutron, but why? We now know that it is because of three quarks, i.e. building blocks.' But these quarks which keep the protons together have slightly different properties, so how these quarks behave and what they themselves are made of is unclear. Aside from these particular questions there is the broader conundrum: how the four forces, electro-magnetism, the weak and the strong force and gravity relate to each other. 'Are there are common themes between them, are they manifestations of one force or not?'

One of Cashmore's current enthusiasms is asking how quarks behave, and how they are held together by another particle known as a gluon. Gluons are known as exchange particles, moving back and forwards between different quarks, acting as a binding agent between the two constituents in a nucleus, the proton and the neutron. Gluons were discovered in an experiment in Hamburg in 1979 and have since excited scientists all over the world. From his experiments Cashmore surmises that the quarks radiate gluons just as electrons radiate photons, though the processes are different.

The significance of the research undertaken by Cashmore and his team is that it reveals a little more about the dynamic behaviour of particles. This kind of work might seem highly esoteric and more productive of further questions than of answers which would have any practical appliactions, yet, as Cashmore points out, the apparently abstruse theories of quantum mechanics which were worked out in the early decades of this century have led to some dramatic inventions.

> The semi-conductor silicon chip is all based on fundamental quantum mechanics which is about electrons moving around and the interactions in the environment created by other atoms. So although the research of today may seem impossibly remote, we have to remember that we live by physics, by the products of yesterday's physics research.

He adds:

> Physics for me isn't just about experiments. It is about the world around me. I am fascinated why one book over there is blue and one over there is red. The answer to that again goes back to quantum theory – it's because of their atomic structure. The different colours have different atomic structures and they absorb different frequencies of light. So I feel that I have a deeper, more wonderful appreciation of things because I don't only know that that book is blue but that it has a different atomic structure from another one.

If scientists can really understand how gluons work then it might have a major bearing on one of the big questions in science: the relationship between the two fundamental forces that operate within the atomic nucleus, the so-called weak interaction and the strong interaction. Cashmore puts this into perspective:

> Einstein attempted to unify two forces – electro-magnetism and gravity. Since the early 1970s we've got a proper understanding of the connection between electro-magnetism and the weak interaction. One of the next excitements that lies ahead is to show the relationship between the weak interaction and the strong interaction, which we think is mediated by gluons.

It is the quest for the answer to such questions that is driving scientists in Europe and the USA to build bigger and bigger accelerators – several miles long in some cases – in which to accelerate particles, do experiments on them and frame new laws of physics.

As a member of the SERC's Nuclear Physics Board, Cashmore is very familiar with the financial pressures to trim back spending. He argues that in percentage terms, particle physics has been taking less and less of the available research funds and certainly should not be cut any more. What particularly worries him, he says, is the unfavourable impression that scientists abroad are beginning to have about the British commitment to physics. He explains:

> People outside Britain have been saying that British particle physics is not being supported. When people feel that is the case, then they start to visit you less. You don't get as many people to do work with you in sabbaticals as you used to, and your own people are not exposed to the same amount of cross-fertilisation that they have had in the past. We still have people coming here from overseas, but this sort of

contact is poorer than when I first started in physics.
He adds: 'People look askance at physics in Britain.'

Cashmore also referred to a phenomenon, which appeared to becoming more serious as this book went to press, which is that scientists are spending a greater proportion of their time just preparing grant applications. Cashmore says that even in Oxford, 'atomic physics is having a hard time getting funded', adding, 'Everyone is running like mad just to stand still, and that's not conducive to doing good research.'

Some physicists feel so strongly about these problems that they have joined the Save British Science group to campaign for science as a whole. One such is Jean-Patrick Connerade. Connerade is an atomic physicist and is Professor of Physics at Imperial College, London. In Connerade's view, what has really changed the position in physics research (as indeed in other branches of science) is that until the demise of the dual support system of funding many projects used to be fundable from within the universities' own financial resources. However, in the last ten years, the amount of money that universities have had available to support research has declined as a proportion of their total income. This has meant that academics like Connerade have become completely dependent on the Research Councils for all the funds to carry out research. Meanwhile the Research Councils, faced with a rising number of applications, have seen their budgets – set by central government – just about keeping pace with inflation, but not keeping up with 'the real cost of science and even falling behind the rise in the Gross Domestic Product'. The result, says Connerade, is that 'every scientist has acquired a lot of administration'. He adds, 'Even the simplest experiment requires many forms to be filled in and committees to meet in order to raise money. You might spend six months in the process.' One result, is that 'it would now be completely impossible to try a new idea on a short time-scale. It is much harder now to open up a new area to research in the face of more flexible funding systems overseas.'

Connerade's own field of research is what he describes as 'the quantum many-body problem'. This is the study of the interactions between several charged particles. The questions raised go back as far as Newton's attempts to solve the classical three-body problem, and the many-body problem itself is a thriving area of investigation. It is one of the important unsolved problems of physics. Connerade explains:

> This research is not just about the structure of matter. It is about dynamics, about how electrons are excited. It is about the rules of the game whenever you put several particles together.

Atoms heavier than hydrogen, for example, have many electrons, which means that they are a many-body system and they therefore should exhibit a special set of effects relating to their many-bodied characteristics. One of the questions that this poses is how these 'effects' can be uncovered and how they can best be described.

This field of research also relates to the basic principles of quantum mechanics and some of the ideas of its founder, Neils Bohr. Bohr had tried earlier this century to quantise each electron (i.e. creating equations describing the orbit of each electron around an atom individually) but another scientist, Langmuir, suggested that they should be quantised collectively. The Bohr theory was simpler, but left questions unresolved. The theory of many-electron atoms has a number of complications, says Connerade:

> One is that many-electron atoms present difficulties not only in their quantisation but also in terms of relativistic dynamics: it was found that if you tried to write the relativistic equation for a many-electron atom then the correlations turn out to be a major problem. Furthermore, if you have just two particles you will have closed orbits. In principle, you have the possibility of chaotic motion.

One of the exciting aspects of this research is that it has a bearing on chemistry. Each element has a different atomic weight, which determines where it fits into what is known as the periodic table. There is a lot of evidence, says Connerade, that 'chemistry not only depends on the one-electron model but also on multi-electron correlation effects: that is, the dynamics which makes chemistry possible may actually be connected with these many-body effects.' He goes on:

> One of the questions that remained after quantum mechanics had more or less been worked out by Bohr was: how does chemistry work? Science had solved the problem of stability but there remained the problem of the dynamics, how chemical reactions take place. Where you have a many-electron system, what is the response of the system to an external disturbance?

> The connection between physics and chemistry may be more complicated than that enshrined in the periodic table of the chemical elements. Indeed Bohr's theory is only true for the first three columns of the table [about half], and the full picture is only now being worked out. Even the periodic table is more complicated than meets the eye.

To try and get at the answers requires some lengthy experiments. One method involves exciting several particles in a high energy field using a laser beam. Another approach is to use pulsed excitations in order to look at very transient responses. To achieve worthwhile results in this kind of research one needs lots of stamina and curiosity as well as a good memory, says Connerade. He explains:

> You need to be able to store mentally lots of information, so that you can recognise a different pattern or see an old pattern in a different context. You also need a reductionist approach. You should be able to see an enormous number of things and realise that they are unnecessarily complicated and out of that to construct a simplicity. In physics generally one aims for the simplest system, and then one tries to set up laws for this system which one would hope have a more general validity.

This search for simplicity has an almost 'metaphysical sort of pleasure', he says:

It involves the idea that because you have found this simplicity or beauty that you
have got close to some kind of metaphysical truth. Physicists are rather wary of this
because metaphysics has been excluded from their domain. Most physicists are
embarrassed by metaphysics but they are excited by being a little closer to the truth.

Indeed Connerade argues that physics has to a large extent supplanted the
position of the traditional metaphysicians and philosophers:

The philosophers have long asked themselves fundamental questions about the
universe but no precise answers have been forthcoming until the questions in some
degree became the domain of science. Some of the greatest discoveries in this field,
e.g. the Big Bang theory of the origin of the universe, have not come from the
philosophers but from the physicists and astronomers.

Like Cashmore, Connerade feels that an understanding of science ought to be
central to anyone living in the twentieth century:

The ability to live in a technical universe is greatly enhanced if you know something
of the principles upon which it is based. If you believe that you are well adapted to a
society in which you live then it should include a level of understanding of how it
works, how it has evolved and how it is moving. If you were living in the Middle Ages
or the Early Renaissance you could live quite well without having any knowledge of
science, but today if you have no knowledge of science it becomes very difficult to
participate in that society.

The attraction of researching in atomic physics says, Connerade, is that whatever
you do has ramifications that are often not foreseen. 'You find that you are linked
to human activities that you did not expect.' Although he expects no immediate
answers from his research into the many-body problem, the excitement is the
unpredictability and the impact that it might have at a later date. He cites the
example of the French physicist, Kastler, who spent several years after 1945 in a
small Parisian laboratory engaged in 'optical pumping', i.e. driving atoms up and
down between two states using radiation. This activity was regarded at the time as
'typically academic' and probably a waste of time, yet it provided the basis for the
invention of the laser.

Like any branch of science, physics is full of controversies; not just about the
science itself, but about the way it should be researched, which areas are the most
fruitful and even the way its theories should be expressed. Sir Denys Wilkinson,
Emeritus Professor of Physics and ex-Vice-Chancellor of Sussex University, says
that he takes a 'a somewhat cautious view' about how certain aspects of the subject
should be discussed. In a paper some years ago he questioned the fashion for
trying to explain the behaviour of atomic nuclei in terms of quarks and gluons – at
a time when quarks and gluons had become a very popular subjects for study. He
felt that physicists should not start thinking in terms of quark terminologies
without 'a thorough understanding' of the past fifty years of nuclear physics.

While he respects the desire to explore the constituents of matter, which is what quarks are currently thought to be, he still retains a sceptical view about the primacy that scientists should attach to quarks in the overall picture of the structure of the nucleus. He explains:

> I accept that the nucleus is made out of quarks but it behaves as if it is made out of neutrons and protons. Until we can find some phenomenon which we can't explain in terms of neutrons and protons and which demands the introduction of quarks we should go on using the old language of physics, otherwise there will be intellectual chaos in the subject. There is no mechanism available at the moment for talking about complex nuclei in quark language, i.e. in relation to quarks.

In professing this view, Sir Denys is touching on an issue common to all sciences, which is the problem of maintaining an accepted language for a subject while at the same time that subject itself expands and fractures into separate sub-disciplines, each resonant with its own vocabulary. He says:

> I am very much in favour of exploring the nature of the language that we might ultimately have to use but I oppose its being actually used in practice until we are forced to do it. I am not doubting the existence of quarks and gluons, but in the context of a complex nucleus we should recognise that they are there, but not force properties upon the nucleus in terms of gluons and quarks which we can readily understand in more conventional language. Some day we will go beyond that language and that will be exciting when one has to recognise that even inside a piece of coal and in you and me there are quarks and gluons but that is not yet.

Sir Denys has long been fascinated by the language in which we try to express some of the contradictory aspects of physics. For example, in describing an electron, scientists have to say that sometimes it behaves like a wave and sometimes like a particle. Thus the analogies that we use in defining a phenomenon survive 'but common sense has had to be abandoned'. But how much longer, he asks, will such analogies continue to be useful as new discoveries emerge? What would we do if our analogies failed us, if we discovered that in order to explain some phenomenon we had to accept that $2+2=5$? We would then be faced with two choices: either to define new concepts that have no relevance to 'the world of our own experience', or 'to stick with our familiar conceptual analogies but invent an irrational logic, an essentially inconsistent mathematics, to manipulate these concepts'.

Sir Denys argues that to have to make either choice would be hard, and that either would impose a limit on our understanding. Yet we may have to learn to live more easily with patent contradictions. Most people and most societies behave in contradictory ways, which is what many psychologists and historians perceive. What is involved here is taking this process much further, that is, coming to terms with contradictory views of reality that are equally valid.

An area of physics that has become important and which promises a somewhat more immediate impact is experimenting with nuclei at high speeds. Some of this work in Britain has been undertaken by a team of scientists led by Professor Sharpey-Schafer, Professor of Physics at Liverpool University. The experiments by Professor Sharpey-Schafer involve spinning nuclei in order to understand the properties of the material inside them. Spinning even ordinary objects, he says, will reveal a lot about them. If you found a stray egg in your refrigerator and wanted to know if it was cooked or not without breaking into it, you could spin it to find out. If it spins rapidly and easily then it is solid inside and cooked but if it spins sluggishly then the insides are fluid. So the same principle can be used in a more sophisticated way to test the material inside other kinds of matter.

The experiments involve breaking pairs of protons or neutrons and putting them in orbits going in the same direction round the nucleus. If the nucleus is deformed then this will alter the relative energies of the individual orbits. The deformed structures are extremely interesting and have a drastic effect on the overall dynamics of the system. Sharpey-Schafer has produced atomic nuclei that take on extraordinarily different shapes from pears to footballs. The shape of many of these nuclei is calculated by examining the gamma rays that they send out.

To the layperson this may all seem so rarified as to have no possible bearing on them or even their children. Not so, says Sharpey-Schafer. He believes that a brain scanning device with a higher resolution than those currently in use could be developed from this research. There is also the possibility of designing devices to detect the presence of certain metals in the human brain such as aluminium, which is regarded as a factor in Alzheimer's disease.

Outside the world of experimental research physics there is the world of applied physics, with the two worlds being connected by a common base of knowledge. Physics is essentially the father of modern engineering and electronics. It has made possible most of the advanced processes and devices that we now take for granted, from aircraft and electrical generators to whole-body scanners, video recorders and microprocessor-controlled domestic gadgets. This applied end of physics has attracted as many if not more scientists as has the more fundamental branch of the subject.

One scientist who has spent most of his life in applied physics – principally optics – is Harold Hopkins, Emeritus Professor of Physics at Reading University. Hopkins invented the television zoom lens, was an early pioneer of fibre optics and was responsible for the modern generation of medical endoscopes.

The first television zoom lens appeared in about 1947. Prior to that time the BBC had tried to do zooming electronically but this did not really work. Hopkins, who had been seconded for war work with a North London microscope company, W. Watson & Sons, was contacted by the BBC's technical director in 1945 and was asked to design an improved zooming device. Hopkins advised that trying to improve electronic methods of zooming would be pointless. Instead he proposed the design of a whole new piece of equipment using lenses. Hopkins had already come across a rudimentary zoom system created by a Polish technician, André Gramazki. This consisted of an ordinary camera lens and three telescopic lenses in front of the camera. The camera lens remained fixed while the middle of the telescopic lens could be moved in order to vary the overall focal length. The pictures it gave were poor and the camera could only be used with very low apertures. It was says, Hopkins, something of a toy. Hopkins, who had been researching the theory of image formation, and had advised the government on aerial reconnaisance photography, decided to try and take Gramazki's idea further.

In order to produce something serviceable, Hopkins had to invent a zooming mechanism that not only gave better pictures but one that could meet a number of needs specific to television. These included the requirement that the zoom effect would be available on both distant and nearby objects 'without any adjustment other than a simple refocusing', and that the aperture of the system, once fixed, had to remain unchanged in order that the brightness of the image would remain constant. Hopkins' final solution was to design a single device composed of five lenses, consisting of one fixed lens at one end, two differently shaped fixed lenses at the other and two moving lenses in between. Hopkins says that the design of this device was achieved after months of 'endless thinking and hard work'. When he had perfected his lens the pictures it gave were better than most of those taken with the BBC's fixed focus lenses. The first full use of this zoom lens was during a Test Match at Lords in about 1948 watched by Hopkins from the roof above the Long Room.

Hopkins' next invention, in 1952/1953, was the first flexible medical endoscope, i.e. a tube which can be passed into the body through which a doctor or surgeon can view tissue. At the time gastroscopes for looking into the stomach were made of steel, about ten or eleven millimetres in diameter, full of lenses and completely rigid so that a patient's body had to be contorted to enable the endoscope to pass down in a straight line into the stomach. They were, says Hopkins, absolutely barbaric and not of much use. Their image quality was poor and their viewing range was limited, for example they could not be used to inspect the roof of the stomach where lesions such as cancer tissue form. Hopkins was approach by a leading gastro-enterologist, Dr Hugh Gainsborough from St George's hospital in London, who asked him if he could design a better version. The result was a fibre-optic gastroscope, i.e. a bundle of fine glass fibres, sheathed in rubber.

The inspiration for his idea came from the Victorian physicist, John Tyndall, who used to amuse evening parties by conducting light through a jet of water. He also showed that a beam of light could be transmitted down a glass tube. Both these experiments influenced Hopkins. He pondered the possibility that a group of glass fibres might conduct light like a glass rod. So he tested his idea by heating a glass rod and drawing it into a fibre. When the fibre had cooled he shone a car lamp through it from end to end. To his delight, the device worked. With another experiment he showed that an image could be transmitted down the glass fibres.

No one at the time foresaw that fibre optics would one day be used for telecommunications. However, John Tyndall, who died in 1893, had prophesied that his funny experiments would one day be useful in communications.

Having published his design in *Nature* in 1954, Hopkins could not persuade any British company to produce a commercial endoscope based on his idea. Nor could he persuade the government department responsible for science at that time, the Department for Scientific and Industrial Research, to provide financial support for a British firm to manufacture it, on the grounds that it could not be seen to be favouring one company over another. So the idea was taken to the USA, where the first such endoscope was manufactured.

Hopkins next invention was the so-called rod-lens endoscope. Again, he was asked by a surgeon to make an instrument, in this case a device for taking colour photographs in order to study tuberculosis in the bladder. The instrument available at the time for looking inside the bladder was a traditional cystoscope. Hopkins bought as many different kinds as he could and examined each one minutely, measuring their image quality and the amount of light that they transmitted. He concluded 'that even using the fastest film, one would need at least thirty times more light than comes down a cystoscope.'

At first Hopkins could think of no way round the problem, until he considered whether he could turn the technology of a cystoscope on its head. A cystoscope consisted of an air-filled tube with various small lenses along its length; but maybe if one designed a tube made from sections of glass with lenses in between made of air then more light could travel through it. This technique became the basis for Hopkins' rod lens system – it produced eighty times more light than a cystoscope. The idea has since achieved great commercial success, but as with the fibre optic endoscope the first models were made outside Britain. The major manufacturers today are in Germany, Japan and the USA.

Despite the great achievements of physics, both in terms of new ideas and in everyday applications, it has become increasingly threatened as a subject for research. Physics has arguably now become the most expensive science, not just in Britain but around the world. Some of the equipment it depends on runs into millions of pounds, while the costs of running some of the major facilities are also high. Britain also contributes to several large-scale international facilities, most notably the CERN centre in Geneva, built round a twelve-mile underground circular tunnel where teams of European scientists experiment with atomic particles at extremely high energies. These international projects now consume about two thirds of Britain's nuclear physics budget, i.e. £58m out of £85m (1991/1992).

In 1991 the SERC had its budget cut by 7%, which took it completely by surprise. Faced with the need to reduce its commitments by £28m, SERC's chairman Sir Mark Richmond needed to act quickly. There were three choices before him: to spread the cuts widely and scale down everything that SERC was funding; to quit the international projects; or to make some big cuts affecting the larger facilities in physics and astronomy. He chose the last option, which produced an outcry since it involved the closure of the world-class Nuclear Structure Facility (NSF) at Daresbury near Liverpool in 1993, and also the withdrawal of our participation in a European venture in Grenoble. These two decisions will save £8m a year. The decisions were painful, Richmond says, but the requirement to find a source of savings rapidly left him with no choice. He says there simply was not the time to examine all SERC's projects in detail, which would have taken months, but which might have resulted in a different decision.

Scientists such as Professor Sharpey-Schafer were appalled at the proposed closure of the NSF centre at Daresbury. He described the decision as 'blind panic', and not determined by the criteria of research quality. He argues that the centre is being closed just at the moment when some of its research is about to yield some interesting applications. These are mostly in medicine and were cited earlier. The NSF centre has also been engaged in monitoring radioactive fall-out, for example after the Chernobyl disaster.

Scientists abroad were equally dismayed. The President of Tokyo University has said that the Daresbury centre had produced results that were 'epoch-making', and that it is possibly 'the most important centre for nuclear physics in the world'. An American physicist, Professor John Schiffer of the Argonne National Laboratory, Illinois, also weighed into the debate saying:

> Closing down the NSF will not only terminate a first rate international centre that is leading the world in nuclear structure studies, but would effectively eradicate Britain from substantial participation in the discipline of nuclear physics.

The whole future of our participation in big research facilities, whether in Britain or overseas, remains a big question. Can we afford to stay in expensive physics, could we leave some of this research to other countries and pick up the information afterwards? What would happen if we discontinued our participation in CERN, for example? Sir Denys Wilkinson's views are unequivocal:

> Our experimenters would be excluded from receiving important results as they emerged. We would have to wait perhaps two years before we learnt of important new results, and during that time other countries which had had access to the information earlier would be able to use the results, possibly in industrial applications. What you have at CERN is the largest concentration of the developing technologies, such as radio frequency, cryogenics, computing. If our people are not there to bring back the state-of-the-art knowledge in these fields where does that

leave our industry? In any case you can't do science by watching it, you can only do it by actively participating in it otherwise you lose touch with what is happening, with the processes of discovery.

10

ASTRONOMERS:
THE COSMIC DETECTIVES

Seeing is an art that must be learnt.
Sir William Herschel, Court Astronomer to George III
and first President of the Royal Astronomical Society.

Herschel – who was probably the earliest Continental immigrant to contribute to British science, having come over from Hanover – represents some of the abiding strengths of British astronomy. He was resourceful, imaginative and successful. He detected the rotation of Saturn's ring, and the time period of its orbit and that of Venus and he threw new light on the Milky Way. His major discovery, that of the planet of Uranus, astounded London society and won him his court appointment. With a royal grant he built a 40-foot reflecting telescope which enabled him to see the sixth satellite around Saturn. Herschel stands in a long line of illustrious British astronomers, which by the time of his death in 1822 included Newton, Flamsteed, Bradley and Halley. This tradition of excellence has been developed century by century so that today British astronomy is still regarded worldwide as being in the top league. The question astronomers in Britain are asking now is: how much longer will it remain there?

Astronomy is usually classified as 'big science,' i.e. it uses large equipment and it is seen as expensive, relying on major land-based and satellite telescopes. Yet the annual expenditure on astronomy, which in 1992 was about £70m, accounts for a pretty small share of national spending on science. Funding comes from the SERC whose budget stands at about £450m. Total funding of all the Research Councils comes to about £1bn. One-third of the astronomy budget goes to international subscriptions which allow British astronomers to use overseas facilities, such as the Clerk Maxwell telescope in Hawaii.

Although Herschel made money out of his research – selling mirrors for reflecting telescopes – few, if any, astronomers today grow rich enough through their work, and any industrial riches which may accrue are often long-term. The real contribution that astronomy makes is threefold. Firstly, mathematics, chemistry and, above all, physics all come together with astronomy. Astronomy provides new information to all three disciplines, and in turn is influenced and stimulated by them. Secondly, it has a cultural value all of its own, with a principal beneficiary being cosmology. Thirdly, it stimulates instrumentation and satellite technology.

The cultural argument is one that is put forward strongly by George Efstathiou of Oxford University, who is one of Britain's youngest professors of astronomy:

> People often ask why we should spend a lot of money on astronomy. The real reason to do it is for its cultural importance and also because we have shown that we are very good at it. I wouldn't justify it on grounds of usefulness, though there are some practical uses. In any case, we shouldn't only judge something by whether it is useful to us in a narrow utilitarian sense. What I do feel is that it is not a very good advertisement for the human race if we can't offer children some answers to some of the big questions about the universe, and if we can't have at least some people thinking about the grand questions. In addition, scientists are actually making progress which is a tremendous achievement, and among those scientists the British are making an important contribution. Relative to the rest of the world we have done extremely well at astronomy, and therefore we ought to support it.

One of the areas where progress has been made, which is of especial interest to Efstathiou, is in discovering how galaxies were formed and how they behave. Twenty years ago we did not really understand galaxies. Today we know infinitely more, but in answering one set of questions many more have been raised. Efstathiou's research has focused on the rotation of galaxies and their structure. The context in which this research on galaxies has taken place was succinctly set out by Professor Martin Rees, Director of the Institute of Astronomy at Cambridge, in a SERC astronomy review paper in 1990. He wrote:

> A crucial discovery is that galaxies are more than they seem – the visible part of the galaxy is embedded in a much more extensive 'halo' of dark matter, which comprises ten times the mass of the detectable stars and gas. The dark matter could be in faint low-mass stars, or in 'dead' remnants of an early generation of massive stars. More speculatively, the dark matter could be some new kinds of neutral particles created in the Big Bang along with the matter and electromagnetic radiation.

Those few lines summarise some of the major questions which intrigue astronomers like Efstathiou, such as the composition of dark matter (which accounts for about 90% of the universe) and the role that it played in the formation of galaxies.

In 1980, Efstathiou went to the University of California, Berkeley, to work with two Americans, Dick Bond and Joe Silk, on research into galaxy formation. One idea they tested was whether the dark matter in galaxies was formed out of neutrinos (a neutrino being a particle with no mass and no electric charge). The conjecture was that there might exist certain neutrinos that were extremely dense and that had a weak electric charge. If Efstathiou, Bond and Silk were proved right then they would have discovered a new particle that 'had important consequences for the distribution of matter in the universe on the very largest scale'. If it could have been shown that even a neutrino-like particle was a major constituent of the universe, the implications would have been highly significant. However, the theory, in its original form, proved not to work very well. On the other hand, says Efstathiou, 'the theory evolved into the cold dark matter theory where the dark matter was not a low mass neutrino but a more massive particle but still weakly interacting with other forces.'

For the last six years, Efstathiou has been working on the cold dark matter theory with colleagues in Britain and associates in the USA. Developing theoretical models of cold dark matter depends on very complicated physics, and it has become a fairly controversial field. But new mathematical models will not provide all the answers to this conundrum. What will take this field much further will be the new 8 to 10 metre aperture telescopes which will enable astronomers like Efstathiou 'to see galaxies forming'.

The study of galaxies, says Efstathiou, leads you by stages to ask where the universe itself came from.

> Galaxies form by gravity from ripples of matter in the universe. . . . As the universe expands and gets older these ripples collapse and that is how we believe galaxies form. So the question of where galaxies come from can be translated into where do the ripples come from, and the answer to that question requires physics that is so new, where energies are so high, that we have very few links from any sort of experiments that we can do on earth.

So in this way, cosmology becomes a way of studying high energy physics 'in a very crude way,' while some of the most speculative ideas in high energy physics relate back to cosmology.

We have reached our present level of understanding about the universe partly by asking the kind of questions that a child asks, says Efstathiou. He adds:

> The very best scientific problems are based on the questions that puzzle children like where did the earth come from, why do stars shine, where does their energy come from. We didn't know in 1900 why stars shone and what was the source of their energy; now we know that it is nuclear energy – burning hydrogen into helium. The simplest questions often have the most profound answers and they push our knowledge of physics.

Indeed it was the habit of asking questions as a child and never getting complete answers that led Efstathiou into science. He says:

> I wanted to know what happens when you push the light switch, I wanted to know exactly what is going on. I wanted to know why there are colours. I used to ask many questions like that but I found that adults couldn't give sensible answers. It was clear from reading books that the explanations were quite complicated.

It was asking a striking question – why do galaxies rotate – that led to Efstathiou's postgraduate thesis on cosmology and the origin of galaxies. It seemed to him that the rotation of galaxies was intimately related to the presence of the halo of dark matter (referred to earlier by Professor Rees). He explains:

> Individual galaxies arose tidily, but as they formed so their neighbouring galaxies spun each other up, and because there is a lot of dark matter around each galaxy they spun each other more than one would have thought.

Currently, Efstathiou's principal interest is exploring further the cold dark matter theory. The outcome of the international research in this field could be profound, he says.

> Suppose that in a few years we discovered that the cold dark matter was made up of a particular particle. It could be a weak interacting particle, only interacting with other similar particles and other material through the weak force and gravity. So these particles would cluster around luminous objects but they wouldn't be able to shine themselves. Because they would have a weak charge they would be difficult to detect. Such a discovery would be of enormous importance. It would help to answer some of the questions that I and other scientists have: I want to know if the universe has to be like it is, could it be different, could it have evolved differently? Is it unique?

> It's by asking these kind of questions that you draw young people into science. You don't draw people into science by saying that science is the basis of technical innovation – which it is – because that's too specialised and too remote when you are at school. But if you can say, these are the big questions we are asking and we are making progress, and you can help us to answer some of the questions, then you are making science come alive.

Efstathiou concludes by saying that the practical benefits of astronomy are often overlooked. One of the most pertinent benefits today, he says, is space physics and satellite technology. 'Many of the techniques that have been used to study the earth's atmosphere were developed by research scientists working on fundamental astronomical research.'

One of the great problems in funding astronomy in Britain is trying to balance our international commitments against the need to nurture small research projects at home. Subscriptions to international projects are essential in astronomy: the big new land-based telescopes have to be built by a consortium of countries because of their cost, and this is true of satellite telescopes too. But our interpretation of the observations from these telescopes has to be done by research teams back in Britain, and these teams are nourished by the rest of the astronomy community. So the two facets of astronomy – observation and analysis – are both essential to the whole. However, whenever there is a squeeze on funds for astronomy – and there has been a squeeze every year since about 1985 – those in charge of the budget have to decide whether to continue with an international subscription or whether to curtail the work of smaller research teams. What usually happens is a bit of both, with commitment to an international project reduced or postponed and other research at home being shelved. Both kinds of decision diminish the effectiveness of British astronomy.

Trying to plan a strategy for astronomy in such an uncertain financial context is extremely difficult, says Efstathiou. Whether cuts are made at home or overseas, there is usually little room for maneouvre. At home, many good projects have been stillborn because it is often not practical to abandon existing work, and when we try to select which international projects we will stay with, what tends to happen is that the decision is governed more by the strength of a specific agreement rather than by its scientific importance.

Some of Efstathiou's concerns are echoed by Arnold Wolfendale, in whose department at Durham University Efstathiou began his career as a postdoctoral student. Wolfendale, Astronomer Royal since 1991, says that funding for astronomy has been cut to the bone in the last seven years. In April 1991 he addressed a meeting in the House of Commons to put the case for astronomy.

Arnold Wolfendale describes astronomy as 'physics in action in an exotic way'. Like Rees and Efstathiou, he sees the universe as a vast laboratory providing a context in which to study physics in a unique fashion. To those accustomed to the idea that astronomers spend their lives looking through telescopes, it will come as a great surprise to learn that Wolfendale's early career – in the 1960s – was largely spent underground. He was studying cosmic rays, at first in the unlikely setting of various London underground stations. But his most exciting research was as a member of a team working deep in an Indian gold mine. The mine went down 7,000 feet and the work lasted six years, by the end of which the team had discovered a new particle – the cosmic ray neutrino.

These particles originate in the upper atmosphere, where they are created by the impact of cosmic rays. Since they have no mass and no electric charge they have enormous penetration, and the faint traces that betray their passage can only be detected by intruments sited deep underground, where they are shielded from the impact of other radiation. As Wolfendale explains:

> If we had built a detector at ground level, we would have got lots of signals but most of them would have been spurious. By using the earth as a shield we were able to filter out virtually all the ordinary particles.

The discovery of the neutrinos was particularly exciting because a rival, US, team working in a South African mine was also in the hunt. The team of which Wolfendale was a member pipped the Americans to the post by a number of weeks.

A related area of science is the study of solar neutrinos – particles born in the deep interior of the sun. One of the questions that currently exercises astronomers is why there are so few solar neutrinos – about a third as many solar neutrinos as theorists predicted.

Yet what was it that propelled two competing teams of scientists in the mid-1960s to spend several years researching cosmic rays and neutrinos in deep mines? The answer, says Wolfendale, is that conventional scientific theory indicated that they existed. The big question was how many would there be. The team of which Wolfendale was a member were able to identify just sixteen interactions although billions of neutrinos are believed to have passed through their apparatus.

The neutrino itself has been an accepted part of particle physics since the 1930s when the scientist Wolfgang Pauli proposed its existence. It was the Italian Nobel prizewinner, Enrico Fermi, who used the Italian word neutrino, which means 'little neutral one' to describe it. Neutrinos are also of great interest to particle physicists such as those working at CERN in Geneva, since it was hoped that they played a major role in the relationship between the strong and the weak nuclear forces, one of the major issues in theoretical physics. Unfortunately those hopes proved to be unfounded.

This early detection work on cosmic ray neutrinos was followed by other research into cosmic rays. This time the research was undertaken above ground. Wolfendale's team used a 300-ton magnet in order to bend the paths of the particles in the magnetic field and then measure their momentum by looking at the curvature. Wolfendale explains:

> We went up to higher energies, made precise measurements of the spectra of energy and measured the charge ratio – positive to negative. We then interpreted the data in terms of high-energy physics: what was going on when a proton struck a nucleus and broke it up into bits. We were among the first to measure this.

Research into cosmic rays was to lead to work in a new field – gamma rays. Gamma rays have begun to attract scientists more and more. Unlike cosmic rays they travel in straight lines and they are very numerous. Studying both cosmic rays and gamma rays led to a further study on the large clouds of gas relatively near the planet. This last piece of research fuelled an interesting controversy – the persistence of comets.

Wolfendale's team used cosmic ray intensity and gamma ray intensity to determine the mass of gas in these clouds. To their surprise they found that it was much less than had been thought. This led to a number of new issues for astronomers. A principal one was the existence of vast numbers of comets 'belonging' to the sun; these comets should have been shaken free by collisions with the dense clouds of gas. Why, then, after four and a half billion years do these comets still exist? Reducing the masses of the clouds suggested one answer. It also sparked of an argument as to whether the dinosaurs had been killed off by the impact or the effects of giant comets passing close to the earth. Wolfendale clearly relished the debate that his research had inspired.

Wolfendale emphasises a point made by Efstathiou, which is that 'astronomy feeds physics'. Furthermore:

> It is a marvellous vehicle for teaching. If you want to teach thermodynamics and you want to discuss heat radiation, you can draw examples from microwave radiation left over from the Big Bang. What happens in particle physics is reflected in what occurs in the active nucleus of a galaxy.

Like Efstathiou, Wolfendale stresses the importance of international collaboration; but he adds that once we commit ourselves to a project, we cannot say to our partners that we want to opt out for a year because funds are tight and then expect to be welcomed back later. Wolfendale is not alone among scientists in suggesting that our international subscriptions be paid out of a separate budget, perhaps by the Foreign Office as happens in France. The thinking behind that proposal is that it would protect our international subscriptions but without cutting British research in the universities. He adds:

> In a difficult year for funding the amount that is left for other research after we've paid our international subscriptions almost disappears, so your teams of good people break up and some leave to work outside Britain.
>
> Astronomy is one of the few areas of world science in which we have held up our heads high for centuries. If you look at our success in bids for access to the planned space experiments by the European Space Agency our skills are highly regarded.

One of the fields of astronomy which Britain pioneered was the study of pulsars. Several years before Wolfendale was studying the clouds of gas in the solar system, a Cambridge professor of radio astronomy, Anthony Hewish, was surveying the gas that streams away from the sun. This is known as solar wind and comes off the sun's surface at one million miles per hour. Hewish was one of the first people to measure this from the ground in 1963/64 – satellites had already made the first measurements. Hewish discovered that you could measure the speed of the solar wind by looking at the associated radio waves. His calculations of its speed were confirmed by the results from the satellites, and added a new feature – that the wind from the solar pole was faster. This early interest in radio waves in space was to lead to one of the most exciting discoveries of the century – the pulsar or pulsating star – and like many scientific discoveries it was unexpected.

Encouraged by the results of his research into the sun's radio waves, Hewish decided in 1963/64 to look at other sources of radio waves, such as radio galaxies and quasars. Both were interesting. Radio galaxies emit radio waves of great power, but only one in a million galaxies is a radio galaxy. So they have an obvious rarity value. Quasars (an acronym for quasi-stellar radio source, coined by the Chinese-American physicist Hong Yee Chin) are also curious. They are very bright and look like stars but they are now thought to be very compact and distant galaxies. Their extraordinary luminosity coupled with apparently immense energy and speed present many puzzles to scientists. Because quasars are thought to be so old – some of them are 9 billion light years away – they are one clue to the origin of the universe.

At about this time a postgraduate student, Margaret Clarke, who was working with a colleague of Hewish, Dr Bruce Elsmore, picked up an odd signal with the radio telescope at Cambridge. While scanning part of the sky, she noticed unusual fluctuations of intensity from some sources.

Hewish decided to chart these 'twinkling effects' or scintillations, which were assumed to be quasars. In order to do this he needed to build a special telescope with a large collecting area. With a grant from the then Science Research Council he designed a four-acre radio telescope with about 2,000 antennae. It was built by a team of six people over two years, one of whom was a new postgraduate student, Jocelyn Bell. Once the telescope was set up its function was to measure the size of quasars. It fell to Bell to operate it, to survey the sky for 'twinkling effects', and to map and analyse everything she saw. What happened in the next six months had a profound effect on Bell and was to change astronomy.

Bell's brief was to record the incidence of quasars and then 'by observing the way the twinkling changed during the year to get an estimate of how compact each one was.' More than 400 quasars were identified. Each day the telescope's chart recorder churned out one hundred pages, representing twenty-four hours of recording. Bell went in each morning to collect the chart rolls and start analysing them. It was extremely arduous. Bell recalls being 'sometimes 1,000 feet behind', in her analyses. But within a very short space of time the telescope was showing up not only quasars but some rather strange and unexpected phenomena.

Bell remembers:

> I'd scan the charts each morning and on one or two occasions I came across an unusual signal. This began to niggle me. I didn't know what it was, so I just called it scruff. It wasn't a quasar, and it wasn't a radio signal.

At first she thought it might be a man-made signal like a badly suppressed car, but she soon had to rule this out because she noticed that the signal occurred four minutes earlier each day. This meant that it was keeping what is called sidereal time, i.e. that it recurred every twenty-three hours and fifty-six minutes. This was very significant to an astronomer, she says, for 'it is the time that it takes for a star to be at a certain point in the sky, to set, and be seen again [in exactly the same place].' When Bell had recorded a number of these phenomena she went and told Hewish.

Hewish's first conclusion was that what they were seeing 'was just unusual twinkling'. He commented that a more detailed look at the phenomena would clear up the mystery. In fact it only added to it. In order to get more detail the chart recorder speed had to be increased. So instead of the chart pen moving at about one inch every few minutes it now moved one inch every few seconds. This had the effect of lengthening the physical recording of any signal. Having devised a completely new way of proceeding the hopes of Hewish and Bell were initially dashed. For a whole month the phenomenon that they had seen simply failed to materialise.

Bell was crestfallen. Hewish's view was that what they had seen must have been a flare star. Then suddenly, a few weeks later, the signal manifested again. The next evening Bell went into the office and sat by the chart recorder and to her relief the signal occurred again.

The signal proved to be pretty weird, as Bell explains:

The pen was recording what looked like a series of flashes at equally spaced intervals. I laid out the chart on the floor and measured the intervals. They were spaced at exactly one and a third seconds.

Sitting alone with this discovery was awesome and emotional. As she recalls:

> This signal was peculiar, exciting and nonsensical all at the same time. A signal that recurs in that precise way seemed obviously man-made, but why would it happen four minutes earlier each time?

There was also a much deeper puzzle: an object that emits such a constant signal must have large reserves of energy and therefore ought to be fairly large in size, yet this signal was coming from a source that appeared to be very small. The question was, how could this be possible?

Again, Bell talked to Hewish who could not believe what he was seeing. He thought it must be some other astronomers playing around with electrical interference. So he contacted as many as he knew to ask them. The answers were negative.

As more of these signals began to be found, Hewish and Bell became worried. Bell explains:

> As well as being intrigued we were also scared. New results in astronomy from other places in the world were rolling in. There were quite a few interesting discoveries at that time. The danger was that we might be suffering from *folie de grandeur*, that we'd overlooked something, and that we might publish and then be proved wrong and look very foolish.

So a lot of effort went into trying to find ways of explaining the phenomena away. Could it be a satellite, or radio waves bouncing off the moon? Could it be faulty equipment? That proposition was easily tested. Two colleagues were asked to replicate the experiment and also found the same type of signal. Hewish comments:

> It was like a detective story. We removed all the possibilities until we had to say, we have a pulsing star. We knew immediately that we had an unusual object.

Yet the mystery remained: what sort of object was this signal coming from? Hewish puzzled over this for weeks. He says:

> It had to be a long way off because we could measure the pulse dispersion coming through the inter-stellar gas. I read up about dwarf stars – a star that ends about the size of the Earth – but it couldn't be that. I talked to Martin Ryle (Cambridge Professor of Astronomy). He was mystified. I consulted Bullard, a geophysicist who suggested something that we had often thought of, that it was a signal from another civilisation.

Bullard told Hewish that if the signal was on a narrow-band radio wavelength, then it had to be an intelligent planet. Hewish checked this and discovered that it was narrow-band. This discovery was scary, as Hewish remembers: 'We thought, what do we do now? How do we publish this?'

For three weeks Hewish and Bell sought to discover whether this signal was coming from a distant civilisation. Hewish used the Doppler effect to measure the signals and proved to his satisfaction that the signals were not coming from another planet. Meanwhile Bell undertook some sightings in another part of the sky to search for another example of the same phenomenon. Quite soon a signal similar to the ones that had been recorded elsewhere showed up. Bell could not believe what she saw and decided to spend the next night following the chart recorder. At 2am on a bitterly cold December night she waited for the signal to reappear but some of the equipment started to freeze over and suddenly stopped. Desperately, Bell fiddled around with it to get it working again and returned to her vigil over the chart recorder pen, holding her breath as it recorded a new pulsar flashing, this time at 1.2 seconds. Four weeks later she found a third and a fourth pulsar. It was clear that these signals could not be coming from extra-terrestial intelligences. 'You wouldn't expect to have four civilisations in different parts of the universe signalling to the Earth in the same way', she says.

The weeks that followed were intensely exciting. The journal *Nature* published the discovery within a matter of days and Hewish gave a lecture in Cambridge to a packed assembly of physics and astronomy professors. At the end of the lecture the meeting broke into a discussion about what the phenomena really were. Hewish had suggested that vibrating neutron stars might be possible sources of the pulses. Towards the end of the meeting another astronomer, Fred Hoyle, pointed out that this could not the source because the vibrations of neutron stars were too rapid. He proposed that the phenomena were 'supernovae remnants' and this ultimately proved to be the case. Super novae are big stars that explode at the end of their life. As the star explodes the core becomes compressed and turns into a neutron star, a minority of which then become pulsars. (The word pulsar was the brainchild of the science correspondent of the *Daily Telegraph*, Anthony Michaelis, and is an acronym of pulsating radio source.)

The implications of this discovery have been immense, for they pose considerable intellectual problems for astronomers and physicists. The properties of pulsars are extreme in every way: extreme electro-magnetism, extreme rotation, extreme gravity and, most of all, extreme density. Scientists still do not understand precisely how the radio waves from a pulsar are produced. A pulsar is so dense, says Bell, that it is like a biro cap weighing 100 tons.

Hewish remains fascinated by the pulsars' incredibly regular radiowave transmissions. A pulsar is so accurate, he says, that you can use it to measure time more precisely than you can with a any man-made clock. This opens up two possibilities. The first is that scientists can use a pulsar to test Einstein's general theory of relativity. Hewish explains:

> When you have a clock in an extremely strong gravitational field as you do in a pulsar you can begin to check and even to modify the theory of general relativity, which is extremely interesting from the point of view of the expanding universe.

This area has become a very active field of scientific research. The second implication of the discovery is that pulsars could be used as navigational tools for space flight outside the solar system. Hewish very early on took out a patent on a 'pulsar navigation system'. He admits he did this as a bit of a joke, but he harbours the dream that one day – long after we are all dead and buried – space travel will operate with pulsar clocks.

One of Hewish's current enthusiasms is studying inter-planetary 'weather'. This has two purposes. Inter-planetary weather can affect satellites and space probes and it can also affect our own climate. A striking example is the magnetic storm that produced an enormous aurora which stretched across the USA and New Mexico in August 1978. This took the US forecasting service by surprise, says Hewish, but his own weather-mapping had detected the storm several days before it reached the Earth's atmosphere. At the time Hewish says that he and his team were not able to forewarn the US forecasting authorities because they had not developed a method of routinely predicting inter-planetary weather. Today, he works in conjunction with the American Space Environment Laboratory.

In recent years, astronomy has been seen as something of a marginal science and has suffered as a result. Out of four major projects proposed by British astronomers in recent years, only two have so far been funded, with the other two remaining on ice. These projects include sharing in new US telescopes in Hawaii and Chile, building a radar telescope on the solar ice cap to observe solar wind, constructing a gravitational radiation observatory and building a new Cambridge radio-telescope. Participation in the US telescopes and the building of the Cambridge telescope has gone ahead.

British astronomy, says Wolfendale, needs to be supported by a long-term rather than a short-term view. As he puts it:

> Astronomy is the shop-window of science. I regard it as an indicator of what we feel about science. If you run down astronomy then I think it shows that you're not really committed to science as an endeavour. You could say let's put all the money from astronomy into applied science. That smacks of a banana republic. You need a long-term view because eventually the information you get from research in astronomy will be of lasting value.

It is a point of view strongly echoed by Hewish. He says:

> If you only follow bits of science because they look useful, you'll do very dull science. You won't get worthwhile discoveries. When Faraday was playing around with wires and magnets, there was no conceivable use for his work. If he'd been told to do useful science, he would have had to turn his attention instead to gas-lighting. Are we saying that we would have liked him to perfect gas-lights instead of sowing the seeds for electrical power?

Efstathiou puts the case for better funding even more strongly.

> If you look at one field that is developing rapidly – observing high red-shift galaxies –

we are leading the field. [A high red-shift galaxy is one from which longer wavelengths of light are received; the more red-shifted the galaxy is the further away it is. So the attraction of these galaxies to astronomers is their extraordinary distance from us and their age. By studying these galaxies we can know more about the origins of the universe.] We are about to discover some of the secrets of how galaxies form in the same way as we have unravelled the evolution of stars in the last twenty years. We can understand the evolution of stars quite accurately now. But if we are to do this properly we need a large telescope or at least have a large share of one.

We've tried to produce a programme to keep us going but it's not a programme that will make us stand out among the competition. We won't be driving the subject because we won't have enough observation time.

As often in science, a squeeze on funds not only affects quantity but also quality, both in the range and depth of research and in the quality of people who will take up posts.

As senior people change posts, so the spirit and direction of a subject can change too. Some scientists believe that this is starting to happen in astronomy, and they cite the proposed merger of the Royal Observatory at Greenwich and the Royal Observatory at Edinburgh under one manager as an example. The new post will be a managerial post and not a scientific one, with the current directors of the two observatories reporting to him.

Critics of this proposal say that these two observatories could lose their scientific edge if they are run according to rigid, management accounting lines. The directors of the two observatories will lose their autonomy, and therefore their freedom to direct their work to the best ends from a scientific point of view. The result of that would be that these posts will not in the future attract the best candidates. As one scientist put it:

In the past those posts have attracted outstanding scientists who had the vision to move astronomy forward, but the chances now of enticing visionary figures to take up those appointments and really influence how this subject develops are low.

Sir Francis Graham-Smith, retired Astronomer Royal and now Vice-President of the Royal Society, has said that 'the whole status of astronomy could suffer as a result of this merger.'

Astronomy is so clearly alive and growing, says Efstathiou, that it would be a tragedy if we allowed our participation in it to decline. 'British astronomy is now on the borderline of losing its edge. We are still the dominant country in it in Europe, but I doubt if we will be in ten years' time.' We often forget that great astronomers in the past like Herschel were supported by the state. If present attitudes towards astronomy continue then Herschel's advice about the need to learn how to see will become almost irrelevant, since the opportunities to see the heavens will be so few.

Part III
WHAT IS TO BE DONE?

11

EXPLOITING SCIENCE FOR INDUSTRY

*The leaders of Japanese business tell me that they send many
of their brightest young scientists to Europe, and particularly
to the UK, to work here and to pick up the latest scientific
developments. It is on that intelligence that they can then
build, creating profitable new products . . . ahead of market
perceptions and thus setting the trend for the future. It is
precisely in that determined and innovative application
of science and related technology that we are lagging today
as a nation.*
Sir Denis Rooke, former chairman of British Gas,
addressing the 1991 conference of the British
Association for the Advancement of Science.

All the scientists I meet, those working at the frontiers of fundamental science in
the universities, academics who are involved in collaborative projects with
industrial partners, or employees working in applied research and development
inside British companies, confirm Sir Denis Rooke's analysis. They all say that
British firms are less ready to exploit existing research or support new research
than are their competitors.

Of course British scientists have a vested interest in seeing their ideas recognised and taken up, and their criticism that most of British industry (outside
chemicals and pharmaceuticals) is slow to put money into research, especially
university-based research, and poor at exploiting it, might be regarded as
somewhat self-serving and maybe narrowly informed. Industrialists might well
ask, what do scientists – especially those working in universities – know of
industry? The answer is that British scientists are better informed about who is
exploiting science than is often imagined. They travel and meet other scientists
abroad and learn which companies are investing in which area of science, and
at home they see that more foreign companies knock on their doors than do
British ones.

In recent years the failure to exploit science has not only had damaging effects
on British industry, but also on science itself. It has given governments an excuse
not to support science sufficiently and it has sapped morale among scientists. It

has also reinforced the divide between fundamental science, which has been accorded its laurels and applied science which has often gone unnoticed. The result has been that the providers of new scientific knowledge and the users have withdrawn into their own domains, rendering communication between the two difficult.

Despite the obvious advances made by British science in this century its representatives are still seen as essentially impractical, ivory-towered people. They are also considered expensive, asking for ever more money to pursue an interest that may or may not yield any benefits. This attitude not only pertains to scientists in British universities but also to scientists in industry. Scientists in British industry are still perceived as the back-room boys – useful only in well circumscribed situations or in an emergency, but otherwise best ignored. In matters of commerce scientists are often seen as naive, usually by the very same people who have done their best to keep scientists ignorant of what really happens in the marketplace.

Since the early 1980s financial pressures have been pushing academics and business people together. The universities have been forced to look far beyond their campuses for funding and have entered into a large number of collaborative agreements with industry in Britain and elsewhere. Meanwhile British industry itself, faced with technolgical obscolescence, has approached universities and research institutes for new ideas. There is now much more openness on both sides and more imaginative joint research and development programmes, but equally there are still major cultural and financial barriers to the effective exploitation of research. As always, there are more initiatives than action, more committees set up to encourage the use of science and technology than individuals prepared to do something. Moreover, as a whole, the national effort is too diffuse and woefully underfunded.

Most of the exploitation of science falls into one of three categories. It may be academic-led, i.e. the initiative to apply the research comes from a university, university department or an individual member of staff; it may be industry-led, i.e. the initiative comes from a company or group of companies; and in some cases it may be a mixture of the two.

Most universities now have some kind of office that helps academics to commercialise their research. One of the earliest of these was established in Cambridge in 1970, when the Wolfson Cambridge Industrial Liaison Unit was set up. Its early focus was on computer-aided design and engineering, but today it is involved in projects as diverse as polymer science and therapeutic antibodies. Essentially, the Unit acts as a broker between companies wanting to work alongside academic research teams and researchers seeking to commercialise their work. Yet it is very much up to the academics themselves whether they use the Unit or bypass it. Initially, it was the Unit's intention that there should be formal contacts in each university department who would liaise between their department and the Unit, but this system 'remains somewhat embryonic', according to the Unit's Deputy Director, Richard Jennings. So the original potential of the Unit – as a means of encouraging the commercialisation of research – has still not been fully realised. Its biggest activity has been advising academics on patents and licensing. Most of these licences are negotiated through a university company, Lynxvale, which is now earning around £700,000 a year from both licensing and the sale of ideas.

Another example is Salford University which was a celebrated victim of the government's education cuts in 1981. Salford's funding from government sources was reduced by more than 40%; this forced it go out and bring in money from industry, in which it has been spectacularly successful. In the process it almost turned itself into a vast contract research and development centre. Other universities admire it for what it has achieved in this respect but add that a university should not orient itself single-mindedly towards industry's existing needs. They say that a university also exists to pioneer, to research into areas beyond industry's current horizons.

A large proportion of academic research is commercialised directly by a department or an individual. Bill Dawes, a lecturer in engineering at the Whittle Laboratory in the Cambridge Engineering Department, has developed, with Professor Denton, advanced software which is principally sold to aerospace and power generation companies. The software which simulates fluid motion in turbine machinery using what Dawes describes as 'computational fluid dynamics' has earned over £1m, the bulk of which has gone to the Cambridge Engineering Department with the rest going to the university company, Lynxvale.

The advantages of this software to aero-engine manufacturers are considerable. A new engine now costs about $1bn to develop, and its design is infinitely more sophisticated than was the case ten years ago. In modern aero-engines air moves through a by-pass fan at about one-and-a-half times the speed of sound, which raises significant problems in the design of the fan blades and their casing. A company using the software can model the airflow mathematically, thus reducing the design time and achieving significant savings – running the software for six hours on a CRAY supercomputer may eliminate the need for a rig test that would normally cost about £250,000.

Dawes sees four benefits that have accrued to his department from developing this software for industry:

> It brings in income to our department, we can achieve financial independence from any one funding agency, we can strengthen our research and it has introduced us to some very difficult and stimulating problems.

Some academics have gone so far as to start their own companies. Des Smith combines being Chairman of Edinburgh Instruments, which he founded in 1970, with a post as Professor of Physics at Heriot-Watt University. The company, which he runs with Heriot-Watt's Professor of Accountancy, John Small, employs forty people, has a turnover of £2.5m and exports 90% of its output. It manufactures lasers, spectrometers, gas sensors and also makes tailor-made instruments for specific customers.

Although Edinburgh Instruments is on the campus of Heriot-Watt University, it is totally separate from Smith's physics department, which he believes is essential. He says:

> If you want to commercialise some research you have to do it outside the university framework because if you do it within it you'll get it wrong. You have to start from an industrial context and even then you'll probably get it wrong first time. You have to cycle the product round to the market and then back to the company and then back to the university and then back to the company and then the customer. My experience is that if you don't have the guts to go round that loop two or three times you won't get a good product.

The company has developed six products which have all repeatedly been improved. Smith claims that the company's gas sensors are now 'starting to beat the Japanese'.

Yet like many small business people, Smith complains that what he cannot do is expand quickly because he has not got access to the resources. He complains:

> I don't have the financial muscle to push our gas sensor business as hard as I'd like by getting the volume up and the price down. What we need is more capital at a cheap rate of interest. In Japan you can get money at three-and-a-half per cent over twenty years.

Most academic scientists, unlike Dawes and Smith, come up with ideas which do not have a ready-made market. Their ideas are either years ahead of their time, as with some of Smith's research on the use of optics in computers, or they are relevant now but need to be shaped into a product requiring a great deal of development. It is here that scientists face the most frustration. An example is Dr Peter Duffett-Smith, an astonomer at Cambridge.

During his research Duffett-Smith designed a portable radio-telescope. This involved building a control station which picked up signals from a local radio station as well as acting as a monitor for his telescope. By taking two measurements he was able to pin-point the position of a star in the sky. Having built this system, he realised that it had a commercial potential. The invention, which Duffett-Smith calls a Cursor, can be used as a radio navigation and tracking system, e.g. for a police force or an airline company. A vehicle can be electronically tagged and by using the system its position can be plotted on a map back at the control station. The device was mentioned on BBC's *Tommorrow's World* which brought in about 800 enquiries. But initially few companies showed any commited interest.

The Cursor was an idea looking for a market. It is not surprising that Duffett-Smith found it difficult to find a company that would build it under licence, for it needed a lot of development to turn it into a saleable product. However, he has been luckier than many other academics in his position, in that he has at last found some interested companies.

The problem here, as in many similar cases is partly a cultural one. John Fisher, Technical Director in the Technology Division of PA Consulting Group, explains:

> Academics don't realise how difficult it is to get good ideas into the marketplace. Our offices are stuffed with good ideas. You may have a clever idea but it may not be in a form that can make money.

He also adds that what most companies are looking for are new products that they can develop quickly in response to a rapidly changing market:

> What you need is to have more companies like ours that can act as a bridge between universities and the market, which know what the market is seeking.

Indeed this is the route that Duffett-Smith eventually took, using one of PA's competitors as an advisor in selecting suitable companies to exploit the invention.

A much older example of an academically-inspired business is Oxford Instruments, founded by Sir Martin Wood in 1959. Like Des Smith, Wood started his company away from his laboratory and he also retained his university post. Wood's original aim was to make high-field magnets for research laboratories around the world. While working in Oxford's Clarendon Laboratory he had helped design and build Britain's first high-field magnets, and the interest of research centres overseas stimulated Wood to start his own company. It was a hazardous undertaking; high-field magnets were a completely new technology and only ten laboratories in the world then had the necessary installations to make use of them. It was not until the 1970s and the advent of magnetic resonance imaging whole-body scanners, which are built around a large high-field magnet, that Wood's company took off. In the meantime Wood had to sink a substantial amount of family capital into the company and he learnt the hard way about running a business. During its growth the company ran up losses in several years. Wood was not well versed in management accounting, he was not able to run the company full-time and he admits that some of his appointments were not terribly good. He now says that one of the best things that he ever did was to bring in an able managing director.

The story of Oxford Instruments is mirrored by many others – usually having less happy endings – which highlights a worry shared by all academics wanting to set up in business: how can you make your own company a success if you have no commercial background? With academics the problem is frequently compounded by the fact that their company is built round a product that is very advanced and which therefore has to make its own market. Most academics do not know where to turn for advice when they start up in business. Their colleagues cannot help nor can their banks. They need financial, marketing and managerial advice, and finding all three to fit a particular market takes a long time – assuming that the company's founders even try.

Some universities run an advisory service for their small businesses. One

example is the St John's Innovation Centre, just outside Cambridge, run by
Walter Herriot, a former manager from the accountancy firm Coopers & Lybrand
Deloitte. Herriot had been advising small firms for several years before taking on
his present appointment and is very familiar with the difficulties of small high-
technology enterprises. He says that the biggest problem for academically-
inspired companies is getting the founders to think of their creation in terms of a
business and not just as a set of technologies.

New companies often benefit from having at least one outsider on the board, to
counteract the over-enthusiasm of the founders and to add some wider managerial
experience. This is especially true of companies that are founded by academics.
One relatively new Cambridge company, Immunology Ltd, has done just that.
This company was co-founded by Dr Alan Munro, a biochemist, who has worked
in the MRC Laboratory of Molecular Biology and in the University's Department
of Pathology where he was formerly deputy head. Immunology Ltd provides
research and development expertise in biopharmaceuticals for drug companies.
Munro is research director and, as a founder, plays an important part in the
running of the company, but he thought it would enhance the strength of the firm
to bring in two experienced non-executive directors, one of whom comes from ICI
and the other from Amersham International. The danger of being a small firm in
such a competitive environment as pharmaceuticals, he says, is that you either
think too small or too big. Having outside people on the board with long
experience of the industry is invaluable.

One problem that is still underestimated is that academics are often unaware of the
commercial potential of their work; they may publish accounts of their research
and give away valuable information for free. Indeed there is a real clash of values
here. The scientist's natural instinct is that new knowledge should be shared for
the benefit of all, which conflicts with the view from those who fund science – in
both government and industry – that hard-won scientific knowledge should not be
made freely available to your economic competitors. Academic institutions and
Research Council centres have begun to face up to this issue in the last five years,
and are encouraging the patenting of new research. Yet some scientists say that the
Research Councils and university departments have still not taken this question
seriously enough. Brian Richards, Chairman of British Biotechnology Group
PLC and Chairman of the Biotechnology Joint Advisory Board, believes that the
timing of patenting advice is crucial. He says:

> Scientists need to recognise that their discoveries have to be protected. In my own
> area we've made sure that scientists have patent advice in the course of their work
> before they publish. It is never too early for a scientist, even in an area that is very
> abstruse, to be aware of possible applications that might come out of his or her work
> and have regard to how it should be protected.

What is required is a much tighter screening process before academic papers are published. Academic papers prepared by researchers in Research Council institutes will normally pass through a director's office, but a director may not necessarily scan a paper with patenting in mind. In universities, the system of screening is much looser with no obligation being placed upon researchers to show their papers to their head of department before publication.

Another of the barriers to the better exploitation of research also comes from within science itself. A damaging feature of science in Britain is that it still reflects some of the snobberies of those who have themselves looked down on science. Those scientists who devote themselves to so-called 'pure' research are regarded as the aristocrats of the profession. Those who dirty their hands with applied research are somehow seen as inferior in status, while those who work as scientists in industry are looked upon as third-class citizens. With one or two exceptions, such as Sir James Black (at ICI and Smith Klein French), industrial scientists do not win Nobel prizes, and they generally have much less status in the scientific community.

Moreover, those scientists who do try to bridge the gulf between academia and industry are often frowned upon by their peers. Jim Dawes says:

> Scientists who commercialise their ideas are seen as having sold out and having let the side down. Some of the scientists in the Cavendish Laboratory were appalled that my department was earning money by selling some of its expertise worldwide.

One route for exploiting science is for companies to sit besides academics, commissioning them to undertake particular pieces of research. Most of the more successful big companies such as ICI, Rolls Royce and BP do just that. They make connections with academic scientists and then agree a brief with them; this may be targeted at some specific need of the company, or it may be more wide-ranging than that, permitting the academic to research broadly a generic area that is of interest to the company.

ICI, for example, has continuing links with a number of universities such as Liverpool, Oxford, Imperial College, London, University of Manchester Institute of Science and Technology and Birmingham. Birmingham is currently the focus of its research into superconducting materials (the materials that will be used for a new generation of silicon chips). The research is 50% funded by the Department of Trade and Industry through its LINK innovation scheme. Neil Alford at ICI's Chemicals and Polymers Division, who is the manager of the project, says that one reason why this particular collaboration is working is that

> . . . it is a genuine partnership between us and a university. It is not just us seeking information from Birmingham; we have a very special knowledge of superconducting materials which nobody else is making, and the university has skills in electronics. So each party is bringing something to the project [the project also

involves one other industrial partner, the Cookson Group].

This emphasises a point made by other scientists in industry and in universities, that companies can only work fruitfully with academic researchers if they have the skills to match.

Another example is Rolls Royce, which has just signed another three-year contract with Sussex University's Thermo-Fluid Mechanics Research Centre (TFMRC). The deal has been signed by Frederick Bayley, Professor of Engineering and the University's Senior Pro-Vice-Chancellor. The work covered by the contract includes

> . . . research into determining the stresses affecting components in the engine which do not directly handle the main throughflow of air as well as the temperatures those components reach during flying.

A further case in point is provided by BP, which over the last five years has brought together scientists from four universities to develop a means of producing the material polyacetylene in films. The material has potential applications in a number of industries including electronics and silicon chip manufacture. Chemists from Durham, Sussex and Edinburgh and physicists from the Cavendish Laboratory, Cambridge, have worked on techniques for the production of polyacetylene. BP's research director, Sir John Cadogan, has commented that this research has involved both fundamental science – which has led to the publication of academic papers – and applied science.

The evidence from these successes suggests that cooperation only takes place on a significant scale if there is either a firm institutional framework to make it happen or there are strong personal contacts. In the three examples above progress was made because there existed well developed personal relationships, between BP and Durham University for instance, and long-standing company/university connections – for example, Frederick Bayley has had links with Rolls Royce for 25 years. Through long experience and from their recruitment these companies know where the scientific expertise is, they know how to exploit it and they are prepared to put in the funds.

One theme that runs through these accounts of collaboration between universities and industry is that the people in the companies have been enterprising in searching out good scientists to work with as well as forging strong relationships. Dr Derek Birchall, shortly before he retired from ICI, explained:

> I've got a lot of fingers in a lot of pies. I'll get an idea that I want to research and then I'll pick someone in a university and I'll pick the best. The secret is knowing a lot of people. You don't wander into a university blind and expect to find somebody and then just go and work with them. You need a dialogue for a long time.
>
> I remember I wanted to measure the effect of aluminium on health. I thought I knew what happened when aluminium passed into a cell, but I wanted to familiarise myself a bit more with the subject. So I went to a meeting and I realised that I was on the right track. Then I read the literature which gave me a feel for which scientists are

in the field, and then I asked around who else was in the field. So I began to see who was involved. Eventually I phoned someone in Liverpool University, and I asked him if he would test an idea I had. Now we are collaborating on a project. That's how it works, and it's the same for my colleagues.

The importance of personal relationships and the quality of technical knowledge in those relationships is underscored by Dr Brian Richards. His company works with a number of universities, principally Oxford. He says:

The process of technical transfer from university to industry rests on there being a sufficiently well developed communication between the two. The relationship needs to be intellectually intimate enough for the transition from basic science to applied work to be seamless.

British Biotechnology is currently developing a vaccine for Aids which, Dr Richards says, grew out of the company's close relationship with scientists led by Dr Alan Kingsman at Oxford. He adds: 'They recognised that their basic research could be the basis for creating vaccines.' In other words, Kingsman and his team were sufficiently well acquainted with Dr Richards and his colleagues to understand the thinking and the strategy of the company. Some scientists in universities complain that their industrial partners keep them at a distance, so they do not know what long-term plans their partners are cherishing. They believe that in those instances industry misses out because an arms-length approach diminishes the potential contribution that an academic could make to a company's technical progress.

Several academic scientists talked of the bureaucracy of companies that inhibits collaboration and the transfer of good ideas. Professor Tom Blundell:

The real problem with technical transfer is that it is very difficult to get companies involved in a project unless the scientist in the company actually speaks to the person in the university who is doing the research. We all face this problem. I've tried several times over the last ten years to get a company interested in the work of a student. In each case the company says it wants to talk to me, not to the scientist, but that's not enough. You can set up the collaborative structure at a high level but in practice the collaboration will only work if the scientists who are going to act together have agreed on what they are going to do.

Blundell argues, like many scientists, that one of the best channels for the transfer of ideas from academia to industry is through what is known as CASE studentships. CASE stands for Co-operative Awards in Science and Engineering, and the CASE studentships are funded by the SERC. Under the scheme a student is supervised by an academic and by someone in a company, and the student will work on a project which involves both parties. The scheme, says Blundell, is 'a very good and inexpensive way of transferring technical ideas to industry.' Professor Richard Joiner, Director of the Leverhulme Centre of Innovative Catalysis at Liverpool university agrees:

The value of CASE is that it involves a low level of interaction and a small financial commitment by the company, a small level of restraint on the academic and provides a regular contact between the university and the company that can be developed and grown.

He adds, 'It's a scheme where we are ahead of many other countries.' What has irritated most scientists is that the administration of the CASE scheme was changed a few years ago, with the responsibilty for managing CASE being switched from the SERC to university departments. At the same time its financial resources have also been reduced. It has recently been decided to revert to the original method of administration, a decision which has been well received, but scientists ask why the scheme was meddled with in the first place.

There remains one essential ingredient for university/industry collaboration which is often overlooked. This is that industry must itself have an adequate research capability if it is to profit from a partnership with academic scientists. This is the stumbling block on which efforts to encourage British firms to collaborate with academic partners often collapse. For British industry's research base is a shadow of what it used to be. As Igor Aleksander, Professor of Electrical Engineering at Imperial College, London explains:

> Twenty years ago most British industrial companies had their own research and development laboratories – often doing pretty fundamental research – in science and engineering. There were about two to three hundred industrial research laboratories then. They've been wound down to such a point that there are very few left.

He adds:

> Companies still support research in universities but the ethos has changed. They won't support long-term research, and that stems from the government's role in this. The Science and Engineering Research Council (which is a key government agency supporting university/industry collaboration) has had to halt the funding of long-term research because it hasn't got the money.

Indeed, the government's insistence that universities should go out and commercialise their research and sell their skills has been deeply flawed. The ostensible aim was to transfer some of the outputs of science to the commercial world for the benefit of the economy. But, as Aleksander says, many companies were not willing to support any venture unless it had an immediate pay-off. The result, from the universities' point of view, is that they have had to bring in funds from foreign companies that were more willing to invest for the long-term.

One of the worries of British scientists, and indeed of some industrialists, has long been that the country is not putting sufficient resources into strategic areas of scientific research and development. They may be regarded as strategic because they represent new and burgeoning areas such as materials or biotechnology, or they may be considered important from an economic or military point of view, as with information technology or telecommunications. These strategic areas are often seen as the seedbed of new industries. They also involve long-term research, i.e. lasting at least five years, often ten and they are as a consequence expensive to pursue. In such circumstances individual universities can only make a very modest contribution and industry, which has been forced to become ever more short-term in its priorities, is just not interested.

One response in these circumstances is for the government to set up a subsidised programme. The famous Alvey Programme which ran from 1983 to 1988 is a major example. Alvey was established to support the research and development of a number of so-called 'enabling technologies' in the field of information technology. The programme brought together 113 companies and 55 universities, collaborating on 250 projects. The biggest industrial participants were GEC, ICL, Plessey, STC and BT which, between them, accounted for 200 of the projects. The total value of the programme was about £700m of which the government contributed half. As with any exercise on this scale, Alvey was not without its critics. It focused on British companies, many of which were regarded as being profitable enough not to need government support, it discouraged collaboration with foreign partners and it yielded fewer applications than expected. It was also criticised by a London Business School report for being 'wrongly focused', especially in the software field.

Alvey never lived down the fact that it became the preserve of big companies, and this discouraged smaller, and perhaps more innovative, firms from collaborating in the programme. Smaller companies in the electronics industry felt that if they involved themselves in the sort of joint pre-competitive research that Alvey was funding they would be giving away important research secrets to their competitors. Another problem was that if academics wanted to apply for research grants from the Alvey Directorate they would only be considered if they went into partnership with a company.

There were also two serious flaws in the way the Alvey programme was conceived. The first was that it did not form part of a wider government strategy for information technology. Indeed what was odd was that Alvey ran flatly against much of the prevailing philosophy of the government. This was shown by the way in which the government was treating Inmos, a state-owned chip manufacturer originally funded by the National Enterprise Board in the 1970s. Inmos was founded as a mass producer of silicon chips and it had one unique product, the Transputer, a novel and extremely powerful computing device. Just at the time when Inmos's fortunes were rising – with growing international recognition of its technical capabilities – the government sold it off to Thorn EMI, which in turn disposed of it about two years later to the French electronics company, Thomson. The second flaw was that Alvey was exclusively concerned with research and development. As one senior source in the electronics industry says: 'It didn't concern itself with the broader perspective of where this research was leading, who would be exploiting it and for what markets.'

Halfway through the five-year programme it came under attack for not deliver-
ing tangible results, a criticism that was both premature and misguided. Some of
that criticism came from within Whitehall, from people who either had never
supported the programme or who now sought to distract attention from their own
mistakes in framing its goals. Not surprisingly, a year before Alvey was due to
finish the government lost its nerve. It had been hoped that it would be followed
up by a further five-year programme at a cost of £1bn – this had been the scenario
outlined by the participants in 1981 – and by this point a considerable number of
Alvey projects were suitable for commercial exploitation. But already the funds
were running out. Brian Oakley, Alvey's Director and Bill Mitchell, then
Chairman of the SERC both argued in favour of a stage-two Alvey. It is indicative
of Britain's priorities that Alvey was wound down in the same year that the
Chancellor of the Exchequer gave away £2bn in tax cuts.

Derek Roberts, who was Technical Director of GEC at the time, says that the
failure to continue the Alvey Programme was a tragedy and 'did considerable
damage because we never reaped the full benefits', but he adds that there has been
one important long-term benefit. 'Alvey did more than anything else to stimulate
very good collaboration between industry and the universities and it stimulated
collaboration between companies in pre-collaborative research.' In practical
terms, he says:

> It enhanced our national capability in some areas of electronics. If you look at
> manufacturing processes, the design of manufacturing systems, we would have been
> a damn sight weaker if we had not had the backing of Alvey.

Alvey has been superseded by other schemes which address a wider range of
technologies. The major schemes are the Joint Framework for Information
Technology (JFIT) and ESPRIT which is a European-funded programme. JFIT,
which was set up in 1989, was intended as a continuation of Alvey but its resources
are much smaller. It is managed by the Information Technology Advisory Board
(ITAB), whose first Chairman from 1989 to 1991 was Nigel Horne, a former
divisional research manager at STC and GEC. Horne says that a lesson of Alvey is
that 'we need to pull technology through into business applications', rather than
assume that technology itself will push new products into the market. As ITAB's
first Chairman he sought to put this principle into practice:

> My view was that wherever we had a research project we should also have a user of
> that project in the same funding regime. So on the one hand a researcher would be
> encouraged to direct his research to the needs of a user and the user would have
> clearly identified goals as to how he would use the research in his business.

Thus the aim is to have the research team working towards a goal that is fully
shared by the commercial user.

Another initiative is LINK, a £210m research and development programme
which brings together universities and industrial partners. This was conceived in
haste, has been slow to get off the ground and has been bedevilled by problems.

Until recently, LINK grants were only given to projects that involved more than one company; as with a similar stipulation in the Alvey Programme this dissuaded a number of companies, especially small ones, from participating for fear of giving away commercial secrets. As a scientist in one company says, 'that requirement was tantamount to giving the keys of my research department to our competitors'. The assessment criteria, set out and supervised by the DTI, have also been seen as an impediment, in particular the stipulation that intellectual property rights should attach to the business and not to the academic institution involved in a given project. The irony here is that it was the government that initially persuaded universities to exploit their intellectual property more vigorously. A further problem is that LINK was intended to benefit small companies rather then large ones, but the kind of research that LINK wanted is not the sort that small companies can afford, even with the 50% grant towards research costs that LINK provides.

Inter-disciplinary research centres (IRCs) have been a more fruitful source of industry/academic cooperation. These were first established in about 1987 and their purpose has been to encourage collaboration across scientific disciplines within universities and between industrial partners. There are now more than 30 IRCs. These include the Sussex Centre for Neuroscience, mentioned in an earlier chapter, and the Liverpool University Centre for Surface Science. The IRCs were funded on the basis of four-year rolling grants which would be renewable after two years, with the expectation that they would run for ten. Most of these IRCs have been successful but their success has become an embarrassment to the SERC which funds them. As one senior member of the SERC explains: 'There is now the realisation that these IRCs tie up a lot of money in advance, which restricts the flexibility of the SERC, year-on-year, to respond to new research needs.' So the IRCs, which were based on an idea from the USA, are now under a cloud because of the shortage of SERC funds. As the funding for each IRC comes up for renewal over the next one to three years, the concern among scientists is that funding will either be aborted or continued in attenuated form. Commenting on the Liverpool Surface Science IRC, Professor Joyner says:

> The administrative mechanisms for the IRCs have been changed in such a way that it'll be difficult for the first tranche of IRCs to have a significant future beyond six years, that is, a future on the same scale and carrying out the level of science originally intended.

The IRCs appear to be suffering in the same way that Alvey suffered, in that an initial enthusiasm is undermined by a lack of long-term financial commitment. The lack of continuity in policy and in funding creates enormous uncertainty and restricts the ultimate benefits to industry as well as to British science. Senior Research Council figures are defending their disenchantment with the IRCs by arguing that some of them have been badly managed. As one senior figure explains:

We now realise that you need a good research director and that you need to clarify the relationship between the department that houses an IRC and the university within which that department lies.

But the reality is that only about two or three IRCs have been genuinely thought to have been mismanaged. To cast doubt on all the others as well is unfair and suggests that the same process of destructive criticism that preceeded the demise of Alvey is now afoot in regard to the IRCs.

One of the results of the Alvey initiative was the setting up of an institution that would foster greater exploitation of science. This was the Centre for the Exploitation of Science and Technology (CEST). CEST began in early 1988 with a fund of £5m, of which £1m came from the government with the balance being donated by a clutch of big companies – BT, IBM, Rolls Royce, Lucas, Shell, British Aerospace, British Gas, Thorn EMI and Jaguar.

So far CEST has mainly acted as a sort of advisory body investigating the scientific needs of various industries. But part of its original brief was to develop new mechanisms for the exploitation of science. Its most recent proposal is that Britain should create a group of research and technology institutions similar to the German Fraunhofer Institutes. These are essentially applied research institutes, staffed by academics and industrial scientists working to industry-inspired goals. There are 35 such institutes in Germany and the German government is currently setting up ten more – all in the eastern half of the country – at a cost of £10m each. Although they are private organisations they receive between 30% and 40% of their funding from the government, mainly to equip their laboratories. The rest comes from industrial companies who pay the institutes for contract research and support their Ph.D. students whom the institutes train.

These institutes serve several functions. They enable scientists and engineers to move from the commercial to the academic sector and back again. They promote the transfer of new ideas from universities out into industry through the training of postgraduate students, and they also stimulate new research in universities. Fraunhofer institutes, positioned at the interface between industry and academia, will often see a potential technical need in industry and will initiate research in a university or maybe within the Fraunhofer system. The proposal by CEST is that there should be a British version based around what it calls Faraday centres. CEST's ideas were initially greeted with considerable interest. However, if these centres were to be effective they would require government funding, as do the Fraunhofer Institutes, and this realisation has effectively dampened support for the idea.

A small pilot scheme costing the Department of Trade and Industry £2m, started in October 1992, using some existing contract research centres. CEST's chairman, Bob Whelan, would have preferred a £30m pilot programme because he has argued that only a larger scheme would be able to test the benefits of the idea. Maybe £30m is a bit high for a trial, but £2m spread over several centres is completely unrealistic. That low level of expenditure will rule out the buying of first-class equipment which is essential for top quality research and the testing of innovative ideas. It is a miserably half-hearted approach towards assessing a practice that our main European competitor has adopted so successfully. The Faraday proposal certainly has not been helped by the fact that it was concieved without any consultation with the SERC which itself had been quietly working on a similar idea – Parnaby centres. The SERC will administer the pilot Faraday scheme but its own research centres will not be participating.

With the experience of Alvey, LINK and the IRCs behind them, many people in universities, industry and the Research Councils are already having second thoughts about creating fully fledged Faraday centres. Tom Blundell, Chairman of the Agricultural and Food Research Council:

> We all agree that there ought to be centres closely involved in academia and industry, but people have overlooked the huge investment that the Fraunhofer institutes entail. We'll do the Faraday idea a disservice if we don't think it through. There's been a terrible problem in the last ten years to go after fashions, and to rush into things without thinking through the implications either in terms of funding or of institutional arrangements.

SERC's chairman, Sir Mark Richmond, adds:

> We need to think through how the Department of Education will be involved as well as the Department of Industry, and we shouldn't underestimate the cost if we are to do this properly. . . . The idea has already been watered down. It is another example of the good old British practice of seeing an excellent idea somewhere else and than adapting it, but we never put money into these things. The next stage is to say, let's get industry to put in the money and then the idea gets stalled.

In addition, Richmond argues, if the Faraday centres actually worked then they would siphon off students and contract research from existing university engineering and science departments.

These comments are given added weight by Nigel Horne, who as a member of the Cabinet's Advisory Council for Science and Technology (ACOST), saw the Faraday idea gather momentum. He says:

> We have a tendency in Britain to cherrypick ideas from other countries. This will only work if you get the whole package right, and consider the environment within which the Fraunhofer institutes work; they work on the basis of long-term commitment from industry. The industrial secondees who work in the institutes know that they can re-enter their companies after three or four years, but in Britain a secondee can't be confident of that. We also forget that the German companies that make these long-term commitments in personnel and in funding research are supported in their efforts by the German financial system. Our financial system is quite different.

Whatever technically-based initiatives governments adopt in trying to facilitate technology transfer, there remain important barriers to the exploitation of research within industry itself. These are essentially cultural and, as Nigel Horne pointed out, financial. In the cultural sphere there are two in particular. Firstly, industry in general still does not make the most of its technical people and, secondly, industry's non-technical managers know so little about technology that they feel embarrassed to discuss technical issues with their own technical experts. Real managerial responsibility tends to rest with non-technical people who shunt their technicians into a siding.

Peter Houzego, who has worked in the UK electronics industry and is now a consultant with PA Consulting Group makes the point like this:

> Scientists and engineers are not allowed to take as much responsibility as they are capable of and they are badly used. A huge tranche of industry does not use scientists and engineers at all when they could. Very few firms will employ people with Ph.Ds. because they feel that very bright people only cause you trouble. The cleverest people who come out of universities are looked on with horror in a lot of engineering and electronics companies.

John Fisher, Technical Director at PA adds:

> If you look at a typical R & D department you find that the people are in a sort of cage deliberately isolated from the rest of management, in particular isolated from the marketing people. I was recently working with a big electronics company where one saw brilliant research people stopped from talking to the marketing people. The result was that they invented things that they hoped the marketing people would like.

A similar view comes from Colin Humphreys, Professor of Materials Science at Cambridge, who says:

> A lot of the scientists whom I meet in industry are terribly dissatisfied. They find that there is a low level of support for their research and that they are not at all involved in policy-making.

Humphreys compares British industry's approach to managing its scientists with that of other countries, such as Japan. He explains:

> I was recently asked to talk to a group of senior people at Hitachi. Round the table were senior management, research people, the works foreman and people from marketing. We discussed their future plans and the impact they would have on their technologies and we had a real debate between the research people and the senior management. In the UK when I've been asked to do a similar exercise there hasn't been that breadth of senior management involvement – and there certainly hasn't been a contribution from marketing managers.

One crucial reason why foreign companies are much better at using their scientists and engineers than are British companies is that their managers and supervisors share a technical language. They all understand something of the technologies that lie behind their products. So a dialogue between the technical and non-technical people is possible. In most British companies that is impossible; the financial director will understand very little about the technology that comes out of his company's research and development department which means that he rules himself out of having any worthwhile debate with the research staff about the merits of a technical investment. (A key factor here is the narrow arts-based education that most non-technical people have had at school.)

The appreciation of technical issues is not just confined to the senior management of many of our foreign competitors, it is also to be found at quite junior levels of management, and indeed below management level. Miles Padgett, a young physicist recently recruited from Cambridge to PA Consulting Group says:

> I can think of four foreign companies whom we've consulted for in the last year, where I've been asked to give a technical presentation and in each case I've been quite surprised at the level of technical understanding because some of the people I've met have been relatively junior in their company hierarchy. If I were meeting their equivalents in a British company I would not usually expect them to be familiar with specific technical issues.

Both Humphreys and Houzego make the point that 'scientists need to know where their company is going if they are to make a useful contribution.' Both refer to poor morale among scientists in industry because they are left so much in the dark about their company's intentions.

The lack of technical knowledge at the top of Britain's companies and the small amount of influence that technical people have in those companies has a number of consequences. It explains, says Humphreys, why 'British companies are not yet as convinced of the value of research as some of their competitors.' It also partly explains, he says, why 'most British companies have not thought out their strategies in the way that other companies around the world have.' Most important of all it means that British firms are not good at managing their research – a view strongly held by industrialist and engineer Sir Robert Malpas, Chairman of the Cookson Group.

Sir Robert addressed some of these issues in a talk to the British Association for the Advancement of Science in 1991. He believes that 'both technologists and business people frequently underestimate the effort, time and therefore the cost needed to take a product to market, and they underestimate the need to test market reaction throughout the development stages.' He also points to a frequent mistaken assumption by senior managers that worthwhile research and development always has to be expensive, forgetting that success can also come from small incremental pieces of research. A critical mistake companies make is to underestimate the cost of developing a product 'beyond the prototype' and into production. When senior managers then realise the extra costs involved they will often abort the project.

While most scientists and engineers would agree with Sir Robert's analysis, they would dissent from his prescription. For he argues that it is the technolgists who should bridge the gap between technology and business. It is the technologists, he says, who 'tend to create over-expectation for their products, their performance, cost and the time necessary for their commercialisation.' They need 'to relate their efforts more closely to such obvious corporate tasks as reducing costs and increasing the value of the products sold'. They also need to spread technical ideas throughout a corporation, so that they become intrinsic to the company.

Sir Robert's views, which reflect those of many a company chairman, beg a great many questions. If scientists and engineers are to think more of the cost benefits of research they need to have the information with which to do this. They need to know, for example, about their company's costing system and the factors that affect it, and they also need to know about the markets that their researched product is being aimed at and the proposed price when it is launched. But how many companies are prepared to give their technical people that kind of financial information? How many companies are prepared to discuss their corporate plans with their technical staff?

One big mistake companies make is to separate rigidly their research and development from their engineering and production. This is why so many projects that originate from research are aborted. It is not that the initial ideas were wrong. What goes wrong in most cases is the way a whole project is costed – a project will be costed in discrete sections and never integrated. This means that when a research-based idea becomes ready for production the costs of production often turn out to be an unwelcome surprise. Nigel Horne says that when he was a divisional research manager at STC with a budget of nearly £300m, he lost count of how many projects he took to the STC board that were turned down. 'The investment needed to go into production was so large that they had to be cancelled. It is misconceived to think that we always need more money for research and development. The major costs come after development.'

What is needed is for production engineers and management accountants to be kept informed of a research project from the very beginning, so that they can calculate the expected production costs in order that the engineers and accountants can influence the embryonic product designs to enhance the manufacture of the product and reduce its production costs. It is that dedication to detail and to integrated project management that has been the basis of Japanese industrial success.

Any discussion of industry's exploitation of science sooner or later turns to the financial obstacles. Technical people both inside and outside industry believe that a key reason why companies mismanage their research and development effort is because they are fearful of adverse City reaction if they commit too much money to it.

Derek Roberts, former Technical Director at GEC and Plessey, says there are six criteria for judging how much money should be spent on research and development. One of these is:

How will the shareholders and City analysts respond to expenditure on R & D, to the detriment of short-term profit, as an indicator of long-term value? [This] has a much more dampening effect on innovation in the UK than appears to be the case, albeit for different reasons, in Germany, France, Japan and the USA.

Dr Colin Humphreys comments: 'UK companies will not look ahead more than two or three years. Other companies, especially Japanese, look ten years ahead. I know of no UK company that is doing research on a ten-year time-scale.' One result, he says, is that tentative collaborative research agreements between UK and foreign companies break down because the UK companies cannot match their prospective partners in commitment.

Nigel Horne, a former colleague of Derek Roberts, believes that the financial environment of industry is the core of the problem. He says:

This is the one issue that we keep avoiding, and it is the one issue that undermines all our initiatives and best endeavours. I've seen this problem at close range as a board member of so many companies. Your shareholders and your banks are a major disincentive to investment. But their actions are dictated by the financial system. What's wrong here is that the banking system and the fiscal system don't encourage risk-taking of any kind.

He continues:

The shareholder pressure that Derek Roberts and I have both seen arises from the cost of money. If shareholders can't get a greater return from the banks than they can get from industry they'll think industry is doing well and stick with it. If a shareholder can get 10% or 11% out of a bank he will expect to get as much from his shares in a company, and if he can't get it from a return on his shares, he will try to get it out of the capital growth in his shares, so the more hype there is on his shares for people to take over the company the better it is for him. It creates all sort of undesirable influences, and it explains why companies don't value their engineers and scientists because in this kind of climate they're not seen as relevant. Everything the technical people talk about is long-term, but if you aren't able to think long-term because of the financial pressures upon you then your engineers and scientists become an embarrassment.

Nigel Horne's views are reflected in the 1991 ACOST report sent to the Prime Minister. In assessing 'the reasons why UK industry does not invest more in R & D and innovation', the ACOST report concludes that the first reason is 'a higher real cost of capital faced by those investing in R & D and innovation compared, for example, with Germany or Japan'. Horne expands the point:

The government has got to think through what the environment has to be to persuade people to take risks. Something has gone patently wrong. We have lost almost all our major players in the electronics industry and the rest of British industry, including the big names, will be lost unless we do something.

Horne's view is shared by Dr Colin Humphreys, who sees leading British companies slipping technically in comparison to Japanese and German competitors. From his vantage point he can observe foreign companies investing in technologies for the next decade in a way that is not mirrored by British companies, with the result that he believes that the major British companies in chemicals and aero-engine manufacture, for example, will simply lose their markets. He says: 'I think British industry is likely to totally collapse over the next decade.' Nigel Horne comments: 'I don't think he is far wrong. You will see some very famous names taken over.' Names such as ICI and Rolls Royce are mentioned by Horne and Humphreys.

So what are some of the measures that need to be taken? Horne has two suggestions:

> I would point to Germany after 1945. At a time of very high inflation the banks were forced to lend to German industry at very low interest rates and guarantee the money for long periods, which enabled industry to take long-term decisions. We have to do something like that.
>
> I also believe that some fiscal way has to be found to overcome some of these problems. I don't think tax incentives for research and development is necessarily the answer, though it might help a bit. Where a tax break would help is in the process and production stage of new development because that is the part of the new product cycle that eats up the money, and that is also where the Japanese have scored over their competitors.

Horne's emphasis on process engineering rather than on product research is also highlighted by John Fisher at PA Consulting Group, and it is noticeable that all the other major management consultancies are now stressing the importance of this.

What all these arguments have in common is that they point to the fact that British companies and the scientists and technologists that seek to serve them are not playing on a level field. Other countries are playing by different rules. This explains why the various recipes that we adopt from time to time never work because they only attack part of the problem. We cherrypick ideas from others, as Nigel Horne and Sir Mark Richmond make plain, but we never think them through and we never put them in a wider context.

Three principal factors affect our exploitation of science. These are: a lack of understanding of how the innovation process works as a whole, an abundance of experts in different fields who do not understand each other and who pursue disparate objectives, and short-termism. What we have tended to do in the past has been to address the first of these, but the conundrum of how we can achieve better technology transfer will only be solved if we also tackle the other two as well.

12

A PROFESSION IN CRISIS

The damage to British science runs very deep. We have had twenty years of systematic and deliberate attacks on science. In the boom time of the mid-1980s science was robbed of its resources, and now in more difficult times there is little money to bail it out. Scientists welcome the appointment of a Minister for Science and his admission that mistakes have been made by government. But it is almost too late. Morale has been damaged very badly, departments have closed and good teams have been broken up. British science has lost ground, and continues to do so.

Paul Davies, former Professor of Theoretical Physics, Newcastle upon Tyne University, now Professor of Physics at Adelaide University.

Paul Davies' comments, made during a recent visit to Britain, sum up the despair felt by most British scientists. Leaving Britain was a wrench, he says, but in a financial climate where he never knew from one year to the next what the future would be for his department, he saw no alternative but to leave and pursue his profession in a country where science is respected and welcomed.

After years of underfunding in the universities and in the research establishments the position of British science has been severely undermined, and morale in the profession could hardly be lower. Although the government has increased slightly the funding of science for the years 1992/1995, this will do nothing to make up for the years of neglect. The extra sums of money involved are still too small. At the same time investment by the private sector in Britain's scientific base has also been reduced. Since 1990 companies have been cutting back on their investment on new capital equipment and on research and development. When economic thrift is forced upon companies the first item to bear the brunt of belt-tightening is expenditure on future product development and its associated requirement for specialist manpower. So scientists are being squeezed by the private sector as well as by the public sector. However, the scientists' criticisms are principally directed against the government.

Sir Eric Ash, Rector of Imperial College, London, has said that the shrinkage in funding for British science has reached critical proportions. He has argued that, 'The funding crisis of the last few years and the shortfall in the funding that is necessary over the next few years is damaging science to a marked degree.' At Liverpool University Professor Richard Joyner says that 'British science is surviving, but at the edge of the precipice'; The prospects for scientists are so bad that Edinburgh's Vice-Chancellor, Sir David Smith, says that he would not want any child of his to go into university science now because there are just so few vacancies and future prospects are appalling.

What has most dismayed the scientists has been the cuts in funding which have affected all branches of science from physics and engineering to medicine and biology. These cuts have often come unexpectedly, halfway through the lifetime of research projects. Thus a project that was intended to run for at least six years can be aborted after using up its first three-year tranche of funding, leaving the team of scientists high and dry. Among the more publicised cuts was the reduction of £28m in 1991 in the total research grant from the SERC, representing 7% of its planned expenditure, which resulted in the decision to close the Daresbury Nuclear Structures Facility and in funding cuts at the world-famous Rutherford Appleton Laboratory. These are just two examples. Every one of the universities and each of the five research councils has also experienced funding problems.

Government ministers have dismissed the protests of scientists by arguing that there has been an increase in science spending during the last ten years. The Prime Minister, John Major, has claimed that by 1994/95 'the value of the science budget will have risen by 31% in real terms since 1979/80'. But the scientists say that it is a misleading statement since it only refers to the monies from the Research Councils. It leaves out the funds that have come from the Universities Funding Council (UFC) and the Polytechnic and Colleges Funding Council (PCFC). The Save British Science (SBS) group argue that, in real terms, there has been a decline in what has been spent on the science base, partly because reductions 'have occurred in the UFC component of the science-base funding'. SBS points out that if rising research costs are taken into account – using a research costs index – it is clear that there has been a decline of 20% in spending.

In July 1992 the Cabinet's advisory body on science, ACOST, broadly endorsed the claim that research costs have risen faster than inflation and came up with a cost index of its own, based on university pay and prices. Indeed it has only recently been recognised that research involves rather heavier costs than had been thought, because the universities' own accounting methods have been under-estimating the overheads involved.

UNPREDICTABILITY IN FUNDING

The financial difficulties of trying to fund science adequately have given rise to a number of issues that now deeply trouble scientists. One problem that has caused widespread exasperation has been the sheer unpredicatbility of science funding. Professor Brian Heaton at Liverpool's chemistry department explains:

> Until 1974 we had five-year contracts with the University Grants Committee. So you could plan for student numbers, and you could close or start departments, knowing that you would be funded for five years. In the 1980s we began to have annual funding, so you could only plan for one year. In the 1990s we have had annual funding and the criteria for funding have become more variable.

What is changing are the ratings for teaching and research. So if in one year a university department retains a high rating for research – and therefore gets more money for it – but gets a lower rating for teaching, the total income of that department may go down, which means that the department has less money to pay salaries. Inevitably, this would affect the research effort. Conversely, a department might receive a higher rating for teaching and be expected to plan for a bigger student intake, but the extra funds for teaching might not be enough to cover the additional cost of salaries. This is happening at Edinburgh University where, says Sir David Smith, there is pressure to increase the student intake and to keep down the numbers of staff in order to make more efficient use of resources. But this means that research will suffer.

In addition to the changes in department ratings, which determine the levels of funding by the Universities Funding Council (which succeeded the University Grants Committee), there are also short-term alterations in research funding by the Research Councils. Professor Annette Dolphin, for example, points to the see-saw in funding by the MRC:

> First there were cuts in funding for projects by the MRC. Now we are told by the MRC that there is a bit more money. You can't fund research properly in those circumstances. It's just an unworkable approach, lowering and raising funding so unexpectedly. How can you plan your research needs if you never know from one year to the next what the science policy is going to be? There's got to be more stable funding.

Sir David Smith also makes the point that research is not something that you can turn on and off like a tap, adding: 'Research is like a plant that you grow; you can't cut it one year and expect to grow it again two years later.'

The frustration at the way science has been handled was well summarised by the irritation shown by Lord Porter, in his outgoing address as President of the Royal Society in December 1990. He said:

> It is difficult for ministers, some with little or no secondary education in science, to appreciate the anger and frustration which scientists have long felt at a system which is guided by those who have little understanding of what makes scientists tick or appreciation of what science has done and will do for mankind.

UNDERMINING CURIOUSITY-DRIVEN RESEARCH

A recurrent feeling among scientists is that the worth of science is just not appreciated. One of the many legacies of Thatcherism was the doubt it threw upon the need for fundamental scientific research – that is, research that probes beyond current knowledge, purely for its own sake and without a clear social or commercial end in view; or, to put it another way, the pursuit of science for the sake of science. For example, Sir David Weatherall expresses amazement that there are still people who ask, 'Why are you spending all this effort on cell biology, why don't you find a cure for back-ache?' He adds, 'Those same people are quite likely to have been given a drug at some time that owes its development to some very fundamental research.' This questioning of basic science in preference to some desired and applied objective has not only demoralised the profession, it has also been misguided from an economic and social point of view. It shows a lack of understanding that science is very often advanced by asking questions without knowing where those questions may lead.

The arguments against a narrow goal-oriented attitude towards research are based on professional as well as practical grounds. To try and put a value on some research before it is undertaken is to question a scientist's creativity, says Sir Hans Kornberg. He feels that it is an implicit questioning of a scientist's credibility and scientific integrity. A even more emotive and heartfelt response comes from Dr Derek Ratcliffe, former Chief Scientist at the Nature Conservancy, who says:

> The primary purpose of science is the pursuit of knowledge, and if it has a utilitarian output that is secondary. If you don't have the love of knowledge for its own sake as your guiding principle then you start to lose your own soul as a scientist.

There is also a more pragmatic argument against restricting science to narrow objectives. If you restrict the paths of research you restrict the outcomes, and as a result you are likely to miss out on important discoveries. Scientific advance tends to move in small stages which cannot be predicted. Moreover each stage involves a restructuring of existing knowledge in order to take account of new information, often from various international sources, to produce a new theory. Even before the age of fast communications, this was the way that science has worked, with scientists drawing in ideas from widely scattered sources, experimenting with novel ideas, drawing a new picture of what they saw and then challenging the accepted wisdom. The experience of science shows that research rarely moves in a straight line from the question to a neat, foreseeable solution.

The experience of science also shows that the utility of a particular discovery is rarely apparent at the time. Most of the main advances in electronics, engineering, medicine and health that we take for granted today are based on a series of incremental and isolated discoveries, each of which appeared to have little practical value at the time it was first made.

Take open-heart surgery as an example. The means by which modern heart surgery is undertaken is based on at least 25 different innovations, each of which is the result of scores of individual scientific advances. These innovations include: electro-cardiograms, blood transfusion equipment, radiological equipment, ana-esthetics, surgical instruments, blood clotting techniques and blood-type label-ling. The two American researchers, Julius Comroe and Robert Dripps, who retraced the scientific pedigree behind open-heart surgery, showed that the development of electro-cardiogram equipment alone depended on 45 separate advances stretching back to 1660, of which 60% were initiated with no clinical intention. The advances were driven by curiosity and they involved issues of fundamental science.

Another example, cited by Dr Peter Collins at The Royal Society, has been the discovery of North Sea oil. This was, in part, based on research by geologists, led by Professor Keith Runcorn at Newcastle University, into the movement of the earth's tectonic plates, i.e. the movement of the continents. At the time this research seemed extraordinarily esoteric, but the long-term benefits have ob-viously been immense.

The questioning of basic science is therefore worrying to scientists. It offends their professional ethos and they feel it compromises their long-term contribu-tion to society. For, although science for science's sake is important to them, it is not their sole reason for practising their skills; most of them hope that what they do will benefit society even if that does not becaome obvious for a generation.

The increasing difficulty in trying to support fundamental research is perhaps best shown by the frequency with which 'alpha-rated' projects are failing to find funding. Research proposals are rated on a scale from the worthy but dull to those that are deemed to have the potential for making a significant contribution to science. The latter category are given an alpha-rating, especially if it can be shown that similar research is not being undertaken in another country. Those proposals that are given an alpha-rating have always been expected to fare better in the scramble for funds than other lower-rated projects. An alpha-rating has never been an automatic passport to obtaining funding approval, but in the past about 60% to 70% of the alpha-rated proposals have secured the necessary funds. Today the average is nearer 55%. It is a clear indication that something is very wrong in the funding of British science. The incidence of alpha-rated projects not being funded is so widespread that every scientist knows of at least one case.

An example was a proposal from the famous Addenbrooks Hospital in Cambridge to develop an animal model to study Alzheimer's disease, work which was said to have already contributed to our understanding of the neuropathologi-cal proteins central to the disease. Despite being given an alpha-rating, the project was rejected by the Medical Research Council at the beginning of 1991 because, it explained, 'funds just do not exist to support this work'. Sarah-Jane Richards, senior researcher at Addenbrooke's, said at the time that the MRC recognised the importance of the work – indeed it had already patented its results – but it could not find her the money to continue the research. In a letter to *The Times* she commented:

> Is it any wonder that a scientist like myself, on a salary equivalent to that of a primary school teacher, on a short-term employment contract and working in a laboratory furnished in part from the hospital waste skip, considers there is no future in science in this country?

At least two Nobel prizewinners have had their applications for research grants turned down since 1989. These were first-rate projects but they did not conform to the set of priorities set out by the Research Councils, which are that research projects must advance a particular field and be seen to be relevant in the national context.

Scientists point out that the rejection of alpha-rated project can affect not only the scientist who puts in the application, but his or her colleagues as well. At Exeter University's chemistry department Professor David Abel talks of a colleague whose alpha-rated project would have involved purchasing magnetic resonance imaging equipment. If the project had been funded the new equipment would have been used for two other pieces of research as well. Abel comments: 'Not getting this machine is a severe blow to my research. I'm studying the fundamental properties of certain molecules, which in the long-term could be very useful.'

SHORT-TERM ATTITUDES

The crisis in science is not just about money; it is also about the direction of research. Jean-Patrick Connerade, Professor of Physics at Imperial College, London, explains:

> The emphasis during the Thatcher years has been on research that gives so-called value for money, i.e. research that will lead to measurable results and which will yield an economic benefit. Blue-sky research has almost been squeezed out. Only a few centres are able to continue with leading edge research. So the research that is being funded is tending to be of the conventional, less risky kind.

This is a point of view shared by Professor Joyner, who says that 'curiosity-driven research is becoming increasingly difficult to support.' Neither the universities nor the Research Councils have enough money for fundamental scientific research. More and more scientists have given up altogether applying to the Research Councils for support.

The pressure to research fairly immediate problems rather than investigate basic questions that may have long-term benefits is apparent in some of the institutes funded by the Agricultural and Food Research Council (AFRC) and the

National Environmental Research Council (NERC). At the agricultural research establishments at Rothamsted in Hertfordshire and Long Ashton outside Bristol there has been a squeeze on long-term research in favour of highly specific inquiries. Fundamental research work is still undertaken at these centres but it is getting harder to justify and harder to fund. The same is true at many other centres of scientific research, for example at the Institutes of Terrestrial Ecology and Freshwater Ecology, both partly funded by the NERC.

One consequence of a targeted approach to scientific research based on immediate market needs is that basic information-gathering receives less and less support. In every field of science, scientists are saying that this rudimentary area of research is withering. For example, the routine measuring of pollution in rivers year in year out (by bodies such as the Institute of Freshwater Ecology) is declining simply because there are insufficient funds to support it. At the IFE's River Laboratory at East Stoke near Wareham, Mike Ladle comments:

> Most of our information is based on work done years ago, because the basis of our scientific information is not being updated at the level we used to do. Therefore it is getting harder to make accurate assessments of the long-term changes in the water quality of rivers.

Professor Richard Joyner is another researcher who stresses the importance of the straightforward recording of new data. He says:

> The nature of chemistry is that it progresses by the accumulation of an enormous amount of facts as well as great advances. But those advances are reliant on bringing together the facts. So the accumulation of facts is an important part of the way the subject advances. It is not especially glamorous but it is important. An example is thermal dynamics, which is a branch of chemistry that tells you if a chemical reaction can occur under certain circumstances; that depends on having tables of heat ouputs. There is no way that that kind of information would get accumulated now.

Innovative basic research is also being crowded out by contracted research for outside clients. This is being forced upon research centres by the funding bodies, both as a matter of policy – to make them more market-oriented – and in response to the government's financial squeeze on science. The trend towards more 'mission-oriented' research still continues, but it is totally misplaced. Professor David Abel points out that university research in Germany, for example, is not mission-oriented, and that German academics are puzzled by our attempt to make it so.

One criterion that has become common in assessing the value of a research project is 'relevance'. The sobriquet of 'relevance' has irritated scientists enormously: the results of most fundamental research have little apparent relevance at the moment that they are made. In any case, what time-scale can be used to assess the value of an idea? Should it be five to ten years, which is what is broadly implied by the relevance criteria? If this is so, are we saying that we care nothing for what society and industry might need in fifteen to twenty years? But, even if one accepted the validity of relevance as a criteria, who is there among us in society who can make accurate forecasts of what kind of scientific outputs will be needed?

Indeed many industrialists say that they do not want university science to be directed towards specific industrial needs. Henry Casimir, the former Research Director at Phillips, made this point in a lecture to The Royal Society in 1987. He said:

> If universities arrange their programmes predominantly by paying close attention to the stated needs of industry they will render poor service to the industries, for it is not reasonable to suppose that captains of industry know what they will need ten or twenty years hence. Universities do not know this either, but by letting themselves be guided by the inner logic of science they may get closer to future needs. In the long run universities may be most useful to industries by not caring too much about their wishes.

Casimir added that he saw no harm in 'industry suggesting certain themes or supporting work it is interested in in a general way', but that is very different from actively directing university research down certain paths. Another point, made frequently in industry, is that if the universities become too oriented towards applied, target-driven research then where will industry turn when it wants to access fundamental curiosity-driven research?

In today's climate a Faraday or a Darwin would hardly stand a chance of getting funded; their work would be seen as far too speculative and they would be told that their research would last too long. Who would fund a Darwin today to join the *Beagle* on its voyage?

British science has thrived in the past because it existed in a creative atmosphere where people were encouraged to pursue their own particular interests. Individuality and unorthodoxy was permitted. It is this quality in British science that has attracted foreign researchers to the UK in spite of the poor pay. Dr Charles Ellington, a young professor of zoology at Cambridge says:

> I came here twenty years ago from America because you had an excellent dual support funding system and because there was more academic freedom here and less bureaucracy. But what do I find now? The dual support system is being wound down and there is much more bureaucracy than there was. So the incentive to stay in the UK is much less.

It is the creativity of British science that has also attracted foreign companies here. Professor Denis Noble says:

> The Japanese companies who have invested in British universities say that what they particularly value is our serendipidity approach to science. To them it is valuable. Why don't we as a nation value it?

THE PEER REVIEW SYSTEM

Some scientists question the whole system by which research is appraised which is based on what is called 'peer review'. An applicant fills in an application form numbering several pages, explaining the reasons for the research and often attaching some further details. This application is then assessed by at least two of his or her peers (some assessors receive as many as thirty applications a month). On the form the assessor must rate the application, placing it in one of five categories, ranging from 'acceptable but not necessary' to 'excellent and likely to advance the subject internationally and should be supported'. The latter are the alpha-rated projects. Increasingly, research applications are also examined by economists in order to judge the likely short-term economic benefit. The chief arguments against this system is that it is said to favour the sort of research that fits in well with the prevailing fashions in any given science, that it distorts science in the interests of some 'objective relevance', and that it does not make sense to taper research projects to fairly immediate economic needs.

Critics of the peer review system are not saying that it should be scrapped; if you have about 10,000 applications a year for research grants you have to have some administrative mechanism for assessing applications. What they are saying is that some of the really interesting proposals are not getting considered because they do not fit in with the criteria, and they add that it is the unusual ideas which will in the long run make the most impact.

One such critic is Dr Don Braben, formerly chief executive of BP Ventures Research which backed scientists with unusual research ideas from about 1980 to 1990. He says:

> Peer review is fine after the event. But that isn't a review, it is preview. If you are that sure of your research that you can state on your application form what it's leading to, then it isn't going to expand the boundaries of knowledge. It is Next Step research, which is important and shouldn't be ridiculed and that kind of research is the foundation of technology, but it won't open up new frontiers.
>
> Peer review works according to specific science disciplines. But the disciplines are a curse, because they make people keep to the beaten path and encourage the view that there is nothing else, and that everything can be reached from where we are now and that if we just keep going on the tracks that we are on now then we will put everything together. But science isn't like that. If you believe, as I do, that we only understand a very tiny percentage of what there is to know, why concentrate in those areas that we understand reasonably well? There's a vast amount of territory that nobody is looking at. But to get to the other parts of the territory you have to think beyond your present classifications and your current assumptions. Peer review doesn't address this, and it never will.

Braben has set up Venture Research International to continue the kind of research funding that he was engaged in at BP. He says:

> We have found a way of finding the people of tomorrow, the people who are asking
> questions that will have a dramatic impact on the way we think, and that impact will
> feed into industrial products. We know from history that this is how it happens; if
> you change the way people think, you generate industrial products.

The peer review system of rating projects before they are funded and the
assessment of progress after they have been funded has rendered the whole
research exercise incredibly bureaucratic.

Scientists feel that they are spending far too much time just filling in forms and
answering simplistic questions. This was one reason that drove Professor Paul
Davies to leave Britain in 1990. He says:

> There is such a miserable attitude towards research, with so much pseudo-
> quantification of research ouput and filling out questionnaires which is ridiculous.
> Scientists are passionate, creative individuals who need to be released from these
> impossible restraints in order to be creative. You wouldn't dream of trapping an
> artist in so much bureaucracy. Truly great works of art are not produced under such
> circumstances, and the same is true of great works of science.

He adds:

> I'm not saying that we should throw money around and let everyone do exactly as
> they please. Of course you need some structure, but the current level of scrutiny and
> assessment works against creative science.

CAN SCIENCE BE 'MANAGED'?

At the heart of this discussion lies the vexed question of 'management'. How far
should we try to manage science? Who should have the responsibility for doing so,
and what should the criteria be? Is there too much management of science from
those who fund science? This subject arouses strong feelings in the scientific
community.

Sir Michael Atiyah, President of the Royal Society, feels quite strongly about
this issue:

> I think a balance needs to be struck here. There will always be some science that
> needs to be pursued because it's regarded as being in the national interest, but
> beyond that I think you need to leave it to the free response of individual scientists
> and to the science that is of the highest standard internationally.
>
> At the moment there are strong pressures to manage science more and more. There
> are lots of reasons for that; if there is a shortage of funds people will say you have to
> manage more, and then you're into managing the decisions on where we put the
> money and that presumes that some scientists or some committees are able to decide
> what is important and that becomes dangerous if pushed too far.
>
> The more scientists are under pressure the more they have to identify what they're
> doing in ways that sound good to the politicians and the outside world. That's a

reasonable response, but if that becomes the main method by which science is defined and by which things can be selected then that can be dangerous. The end result is that the amount of science that scientists are allowed to ferret out for themselves shrinks. That isn't healthy in the long run.

Sir Michael also points out the problems inherent in funding research for very short periods and expecting results over a short time-scale:

The fact that so much science is now funded on a short-term basis means that you have to show results very fast or you won't be allowed further funding. That limits the kind of speculative work that scientists can do and it is a major problem. We are almost seeing funding on an annual basis, we're getting close to that, having to justify yourself in a continuous way. That is bad. Management from the top is not the way to get the best science.

One of those at the top who has talked about the need for more management is Sir David Phillips, a biophysicist and Chairman of the Advisory Board for the Research Councils. In 1988 Sir David addressed the Royal Society on the subject of 'Funding UK Science: Modes of Support'. In his lecture he said there was a

. . .need for a much broader level of management of the resources available, to ensure that they are used as cost-effectively as possible and that the portfolio of research they support is well-matched to the scientific opportunities of the moment and to the nation's needs.

He added that:

The Research Councils have an important part to play, both in pulling together coherent programmes of research, which are comprehensive and avoid overlap, and in stimulating research efforts in fields where more needs to be done in the national scientific interest.

What did he mean by a broader level of management? He explains:

I was referring to the management of funds, the management of time and the management of people, deploying technicians and equipment to the best effect. There are management issues in managing a research lab. The word management always annoys academics. Academics start with the premise that research can't be managed because people mustn't be told what to do but that leads to a blanket condemnation of management.

We need to think about the management of science at five levels. At the individual level it is ludicrous to keep on supporting an individual to do research who has run out of steam and doesn't really know what to do. It is a responsibility of somebody that the person is either moved to another job or put into collaboration with another researcher to broaden his field and move in a new direction. The second level is managing the resources of the lab, managing the equipment and technicians so that everybody is looked after. The third level of management is at the university level: how are the university's resources allocated between different departments. The fourth level is the Universities Funding Council: how are its resources allocated between different universities. Finally, at the Research Councils level we have to ask, what national resources do we need, what major facilities do we need in astronomy or physics? I'm arguing that there needs to be attention to all those levels.

In principle most scientists would agree with much of what Sir David is saying, but they take issue with some of the sentiments that lie behind his arguments. Sir Michael Atiyah:

> Management in a lab has to be of the highest standards, a good manager ought to be able to identify the really good people and then he should let them get on with their research. He shouldn't tell them what to do.
>
> You have to leave it up to the scientist him or herself to decide when they should stop a piece of research. When I work on a problem I will go and worry at it for some time, and then leave it and then I may come back to it when I have got some new insight. Only I can decide when to give up. Research isn't like an exam paper where you have three hours and then you stop. That's a crazy approach in science.
>
> There are people who spend their lives grappling with what seem insoluble problems. Some will never succeed but some will. If you have a cut-off point after which no more time is allowed you will stop those few who will succeed even though they've been unsuccessful for five or ten years. In managing people in a lab. you want the lightest rein possible. The best people who managed laboratories in the past, for example in Cambridge, did it exactly in that way.

One of those Sir Michael has in mind was Max Perutz – the Nobel prizewinner mentioned in an earlier chapter – who managed the famous Laboratory of Molecular Biology. Perutz himself says:

> My laboratory is often held up as a model of a centre of excellence, but this is not because I ever 'managed' it. I tried to attract talented people by giving them independence, listening to them and taking an interest in their work, helping them to get what they needed for it and making sure that they got the credit for it afterwards. Had I tried to direct people's work, the mediocrities would have stayed and the talented ones would have left. The laboratory was never mission-oriented.

Someone who knew Perutz from those days was the late Dr Peter Mitchell, himself a Nobel prizewinner. He said:

> Perutz was very liberal-minded. He believed that the way to get good imaginative science was to choose people like racehorses. You chose those with the best form, encouraged them and then let them get on with their research in their own way. That approach worked.

THE DETERIORATION OF THE INFRASTRUCTURE

One crucial problem facing British science is the sharply declining budget for new equipment and the insufficient technical support to manage equipment. During 1991, the total budget for new equipment in British university chemistry departments, for example, is said to have been around £900,000. One good mass spectrometer alone costs about £250,000. In practice equipment purchases were frozen in 1991.

The impact of reducing equipment and technical support for research is illustrated by two examples. At University College, London, the Provost, Derek Roberts, cites the Cell Development Department which is struggling to maintain the quality of its research with old equipment. He says:

> This department is one of the two largest in the country. It has always got the highest rating for its research. It has eighteen professors of whom six are Fellows of the Royal Society. Yet they are having to work with equipment that is fifteen years old. We desperately need £250,000 this year and next year just to keep the department competitive. Whether we are successful in getting that kind of money will depend totally on securing money from charities.

At Exeter's physics department, Dr William Vennart, who specialises in medical physics, has been trying to secure funding to buy two new computers and to rent a whole NMR imaging system. The department has been working closely with a local hospital, helping doctors to make better diagnoses, of arthritis for example, by using imaging equipment. The new pieces of equipment are required to enable Vennart and his colleagues 'to do better three-dimensional imaging', and 'to do higher resolution imaging of knee-joints'. The total cost, spread over three years, would be about £150,000. If the money cannot be obtained, then this research will not go ahead. This example may seem small, but the morale-sapping effect of not getting the money – which seems likely in this case – is typical of what is happening to research departments all over Britain. The cumulative effect of a series of underfunded equipment allowances over a period of years is what damages the enthusiasm and the output of scientists' research departments. Vennart adds that the problem of trying to get money to purchase equipment for science departments is getting worse.

At the Long Ashton Agricultural Research Centre the Director, Peter Shewry, points out:

> We haven't been able to buy any important equipment for years, for example in mass spectrometry. What is happening is that we are existing by repairing the old equipment, but there will come a time when it will no longer be worthwhile to repair, and we will need to buy a new machine. A replacement will cost about £250,000; I don't see how we are going to get through our budget.

The importance of good equipment is underlined by Sir Hans Kornberg:

> The ability to do good research depends not just on bright people but also on the people having the equipment and the facilities to advance their subject with excitement and high morale. That's the bit that is missing.

Professor Richard Joyner concurs:

> A good part of chemistry now relies on the ability to access very sophisticated equipment like NMR machines, X-ray diffraction machines, high-resolution mass spectrometers.

One of Joyner's colleagues, Dr Brian Heaton, adds:

The success of a project depends on a good laboratory. What concerns me is that in future the universities will not have any money to underwrite our equipment overheads as they have done in the past. This is going to be a very big problem in the future.

In the view of Dr Vennart the problem is already evident:

The mode of support is changing quickly. The financial support that many departments used to get from their universities to buy equipment is not coming from the universities any more. It is having to come from the Research Councils.

The reason that funds are not coming from the universities to the same degree as in the past is that part of the block grant to the universities has been transferred to the Research Councils. The criteria for granting money for equipment is thus changing very fast. Broadly speaking, a university will provide cash for one of its departments for equipment if it has the money and if it believes that the department is deserving. The university authorities will have personal knowledge of the department, and a decision will be fairly rapid. A Research Council operates differently. To secure money from a Research Council, which in most cases is situated many miles away, a department must fill in a detailed application form and satisfy the Council that the research for which the equipment is required is important. Moreover, a department will only have a chance of receiving the money if its research is already highly rated by the Research Council. The result, as Dr Vennart and other scientists point out, is that 'certain departments are no longer getting the money for equipment that they used to'. If departments cannot get equipment, they start to wind down their research, which is exactly what is happening in many universities.

As well as the problem of funding new equipment, there is also the issue of maintaining the quality of the whole infrastructure. At Edinburgh University Sir David Smith complains that 'working conditions are declining because we cannot maintain the buildings as we would like to.'

TIME PRESSURES

The pressure on resources means that scientists are spending an inordinate amount of time just submitting applications for the funds required for research projects and to finance the purchase of equipment. Dr Peter Rich, Chairman and Research Director of the Glynn Research Institute, describes this as the number one problem: 'Getting funds takes up a terrible amount of time.' It does not take hours or days but weeks to put together a decent research proposal. Professor Annette Dolphin says that 'it is almost a full-time job securing funds' for her team of researchers at the Royal Free Hospital. She says that a major research proposal can take the equivalent of six weeks to put together. This is because the proposal itself has to be researched; the importance of the subject matter has to be set out, material that has been published in the field has to be cited, research methods outlined and possible results and benefits listed.

In addition to applying for government funds scientists are often under pressure to bring in funding from clients, be they government departments or industrial companies. Since the mid-1980s, many scientists have found that they are spending about 20% of their time searching for external finance. This problem is particularly acute in those institutes that are largely funded by the AFRC and the NERC, because the government funding does not cover all their costs.

LOSING THE SEEDCORN

The tightness of funding has had a devastating effect on job security and job prospects. With short-term research projects go short-term employment contracts. The notion that most academic scientists have jobs for life is now a thing of the past. Sir David Smith says that at Edinburgh University there are more than 500 staff on short-term contracts (i.e. three to five years), and he wonders how many of them will be able to renew their contracts when they come to an end. He adds: 'I know ex-postdoctoral students of mine who are married with mortgages and children, and they are in mid-career and still on short-term contracts. The job prospects are dreadful.'

For a young scientist in his or her late 20s or early 30s there is, in practice, no structured career path. The opportunity to settle down in a permanent post in one field is extremely limited. Salary levels for those on short-term or – if they are lucky – in permanent posts are poor. The stipend of a postdoctoral student is about £4,500.

As funding for research projects becomes ever harder to obtain, and as career prospects grow worse, there is a flight of younger talent out of science. Young scientists see the financial pressures, the long hours of work for ridiculously low pay, and lack of freedom to research new areas, and they leave. David Eisner, Professor of Physiology at Liverpool university, explains:

> It is very hard now to persuade bright young people to go into scientific research as a career, because in real terms academic salaries are declining and because the money to support research becomes harder and harder to get. So young people feel that no matter how fine their ideas are, they won't get the money to research them. What are they supposed to do? It is inevitable that they should seek to quit science.

The loss of this 'seedcorn talent', nationally and across every branch of science is rather worrying, says Professor Joyner. He goes on:

> The really exciting science is largely done by people between the age of 25 and 40. It is the infusion of new ideas from younger scientists that makes science advance. One reason why we've been so successful at science in the past is that we have given young people their heads.

This issue relates back to the way science is now funded. In the past university departments had research budgets of their own, provided by the University Grants Committee, which they were able to use as they thought best. The money was allocated by the head of department to those individuals whom he or she thought most deserving. No one outside the department asked how this money was spent, which meant that department heads were free to support clever young scientists to pursue whatever they wished. Today, it is much harder to receive support for such 'unhypothecated' research, unless a scientist has a track record or the research is seen as 'relevant' in an academic sense, or is economically useful.

Not surprisingly, the number of applications for research posts has fallen. Moreover, the diminishing number of research appointments has not resulted in a higher quality among the applicants which is what some people – especially in Whitehall – expected. The reverse has happened. Every university science department and every research institute is complaining that the overall quality of applicants is declining. Professor Richard Andrew, a biologist at Sussex University, confirms this:

> There is a general feeling in the universities, certainly among the biologists, that we're not getting the grade of postgraduate applicants that we used to – the independence of thought isn't there and the motivation isn't always there.

Although the best research centres are still able to attract the kind of postgraduate that they want, even they find that the choice of candidates is far more limited than it used to be. Sir David Smith says that it is noticeable that 'major research groups no longer get floods of first-class applicants as they used to'. Centres of excellence like Oxford and Cambridge, Imperial College etc. also have difficulty in persuading some of their own pupils to stay on. The sort of people who walk away with first class honours can see the writing on the wall, and they leave.

Many senior scientists are disheartened by seeing their ablest people leave the science profession altogether. Sir Hans Kornberg at Cambridge says:

> The people in my Part 2 class in biochemistry are the people whose knowledge has been honed to the finest tuning with the expectation that it is they who will be the biochemists in research in the future. But half of them become accountants or actuaries; they may become better accountants, but it's an awful waste. They see no future in science.

Leading scientists have put these points to ministers and civil servants and they have stressed the need to increase postgraduate salaries, but the response has been unsympathetic. As Professor Andrew reports:

> The Treasury asks us, 'Are you returning more studentships unfilled than you used to?' and the answer is no. So the Treasury replies that this suggests that we are paying the right market rate to attract people. But that isn't the issue. We are concerned that the high-quality young people are leaving science.

Many young scientists have quit their profession to start careers in accountancy or financial services. Professor David Abel at Exeter says that leading accountancy firms regard scientists, especially chemists, as one of the best sources of talent.

They are highly numerate and well-organised, ideal attributes for an accountant. In the late 1980s, the City hired a higher than usual number of scientists. One such is Anne Simon, a former astronomer at Cambridge. She explains:

> I was in the second year of my three-year research project, working horrendous hours as everyone does in research, and felt that I'd had enough. I had no security of tenure, I was spending more and more time on administrative problems, and I had less and less time for what I was supposed to be doing and for what I loved which was research.

She looked around, she says, for a career that would offer an intellectual challenge 'where I could solve interesting problems' and which would reward her well. She chose a firm of stockbrokers, Cazenove, where she now assesses high-technology companies for investment.

If Anne Simon had come out of one of the second- or third-tier research departments in UK astronomy, then some people could write off her departure from science as being not particularly significant. But Cambridge is one of Britain's top centres in astronomy. Together with Manchester, it virtually led British astronomy in the post-war era, and it still has an extremely high reputation.

THE BRAIN DRAIN

While Simon is part of a brain drain that is internal to Britain, there is another more talked-about brain drain – those scientists who leave Britain for other countries, usually the USA. This phenomenon is so large that it has led to the setting up of a body known as British Scientists Abroad, which acts both as an expatriate club and as a pressure group for British science. It is administered by Jordan Raff, a biochemist, now working at the University of California. Raff says that British Scientists Abroad has about 2,000 members, including some of the country's most distinguished and senior scientists. By and large, he says, they have left not for more pay, but because working conditions are better overseas and because they feel that science has been systematically undervalued in Britain.

One such scientist is Professor Paul Davies. Davies is a mathematical physicist and also one of the most successful popularisers of science. He justifies his decision to leave in this way:

> I became sick and tired of the uncertainty, never being able to plan for the future, not knowing how much funding I would have in the following year and worrying about how to get hold of new equipment. It wasn't a question of the civil servants saying, 'we haven't enough money, so therefore let us sit down and talk about how we can work out a plan with less resources.' The attitude was rather, 'we think you scientists are surplus to requirement.' Many scientists felt, as I did, that we were just not wanted in our own country.

As with many controversial subjects, assessing the impact of the brain drain is beset with statistical difficulties. Government ministers have argued that the number of scientists leaving Britain is roughly the same as those coming here to work. There are two problems with this argument. The first is that there is no simple cut-off point for emigrants and immigrants in British science. Some people work overseas for a few years and then come back, but others stay to take up permanent posts; while some scientists from abroad come here, sometimes staying but usually returning. So the picture is hardly clear-cut. Moreover, the statistics only take account of those occupying tenured posts; they leave out departing postdoctoral students – traditionally, the young whizzkids of science. The second issue is quality. In this two-way flow of people what is important is whether we are losing our best scientists, rather than the actual total of migrants. The fear is that the alleged small number of emigrants includes a high proportion of world-class scientists who will be lost to Britain for ever.

The Royal Society has already conducted one survey on the brain drain and is in the course of another. Its earlier study, spanning the years 1975 to 1985 and based on a sample of disciplines and institutions, showed that more than twice as many recent Ph.D.s left Britain as entered, 314 as opposed to 115. Half of those who left took up long-term jobs. The total university staff – in the sample – leaving the country amounted to 740, which was balanced by 556 coming into Britain. While 63% of those leaving did so in order to take up long-term appointments, only 35% of those coming to Britain settled into long-term posts. More recent figures from the Universities Statistical Record, covering the period 1980 to 1987, have suggested that the flow of scientists in and out of Britain has become more balanced. But these have been disputed, and in any case are now out of date. Hence the new survey by the Royal Society.

The Society's President, Sir Michael Atiyah, argues that lack of definitive figures for the last few years should not make us complacent:

> What worries me is that by the time you can unambiguously show that there has been a big drain it will be ten years too late. You'll realise that a generation has gone. By the time you have the statistics, it'll all be over.

Sir Michael's current worry is that as the Single European Market gets under way we shall lose many more scientists to Continental countries. He explains:

> It used to be said that people didn't get appointments in other European countries if they didn't speak the language. But the barriers of language are decreasing. British people are now being appointed even if they don't speak the language – this is happening in Germany and Eastern Europe. This may even begin to happen in France. The French are expanding their universities at quite a pace – how will they fill them? They'll have to recruit to some extent from outside. This drain of talent to the rest of Europe is not yet happening in a big way, but I do see the beginnings of it. I see some of my friends and my friends' children choosing to work in Europe. The message is quite obvious: the opportunities are already there, so if there is a serious differential in pay and conditions between Britain and the rest of Europe people will

leave this country.

The brain drain does not only have implications for Britain's science base here and now through the loss of good ideas either to science itself or to industry. It also threatens the future of science. It is now well-known in the scientific community that there will not be enough people in place to fill senior posts as they become vacant over the next ten years. As Professor Kornberg says:

> It is patently obvious. We can predict how many science jobs will have to be filled by the end of the decade, and we know already that there simply will not be enough people to fill those posts.

The succession problem in chemistry, for example, is severe according to Professor Abel. He says:

> There was a big intake of people from 1959 to 1970, but they are now starting to retire. The number of people required in seven to eight years is mind-boggling. We'll be appointing lecturers if they breathe.

Some university departments are already weighted towards those of retiring age: it is said that in the Leeds university chemistry department the average age is 58. Professor Heaton at Liverpool comments: 'What makes the situation worse is that in our current circumstances we can't plan for succession because we don't have five-year budgets any more.'

LOSS OF INTERNATIONAL COMPETITIVENESS

All these factors are leading to one inexorable result: the decline of British science. Some subjects are withering like drought-stricken plants. Dr Charles Ellington, Cambridge Professor of Zoology, says:

> My subject in the UK is dying. There are only about five centres in the UK that are in my area – animal physiology. Here at Cambridge, research proposals are down by two-thirds, and about one third of the alpha-rated projects are not being funded at all.

In the field of physics, one casualty has been research into synchroton radiation lithography (SRL) which is of geat potential importance for the manufacture of submicron-scale silicon chips. Professor Jean-Patrick Connerade explains:

> It is now too late to join this particular field, but the UK had all the right scientific expertise to become fully involved. Our industrialists lacked the vision and our funding agencies the resources. As a result, one or two companies in the UK will end up building some of the tools to be used in the research programmes pursued overseas. We will make something of it, but most of the profit will go to others, which is only fair since we were not willing to take the risks.

Most scientists say that they see that British science slipping internationally. Sir Hans Kornberg:

> In biology our international position is not as good as it was five to ten years ago. If you measure our standing by the number of times that British scientific papers are

quoted, or the patents that have accrued, or the Nobel prizes that we have won, then all these indicators show that we are no longer as pre-eminent as we were. We come second to the USA and we're beginning to come second to Germany and France. It is a view reflected by those at the head of the Research Councils. Sir Mark Richmond, Chairman of the SERC, says: 'If you go round the unversity system, the number of universities that can keep good research going in a range of disciplines is decreasing all the time.'

The comparative decline has become extremely demoralising as scientists see more and more evidence of it. Sir David Smith explains:

Scientists go to international conferences and they learn what equipment their competitors are using – that it is more advanced than ours – and they become preoccupied that their competitors are overtaking them.

This realisation that scientists in other countries are better equipped than ours has a massive impact on morale. Smith adds, 'The government doesn't see the efffect of all this on people who want to be leaders.' It is a view strongly echoed by Professor Dolphin:

What is so sad is that we used to lead the field, and we could still lead the field if we had the resources behind us. We do produce very able scientists, so it seems a tragedy not to support them.

She adds that it is essential for scientists that they can do research that is recognised internationally:

If you were doing science that was only recognised in Britain and ignored outside, then to my mind it would be pointless. It has to be recognised internationally for it to be recognised at all.

A frequent argument from politicians is that Britain should no longer try to be first in science and that it could surely buy in the fruits of advanced research done by other countries. Why duplicate what others are doing, especially when it is expensive to do so? It is a point of view that is often expressed in connection with physics research, which has become very costly to fund. The answer that people like Roger Cashmore, Professor of Physics at Oxford, give is that if you are not abreast with the latest research yourself then you will not be able to understand the newest findings of others. Leslie Iversen, director of UK research at Merck, Sharpe & Dohme – the US pharmaceutical company – believes that a *laissez faire* attitude towards science is both reckless and naive. He says, 'If the Japanese or the Germans are first in a field, they will be the first to use their knowledge. They are not going to give it to their competitors even at a price.'

Politicians argue that a few years of belt-tightening in science research does not matter; the ground can be made up later. Not so, say the scientists. In the world of fast-moving research it is fatally easy to fall behind and never catch up. If you fall too far behind, international colleagues are less likely to share with you their own research, and even when they do so, you may not be in a position to build on that research yourself.

To some scientists such as Professor Davies, these utilitarian arguments in favour of science spending should not be the only ones. He says:

> Because science has economic benefits, it is now being justified almost exclusively on those grounds. The fact that science might stir us with an inspiring world view has been brushed aside as unimportant.

13

SEVEN ISSUES FOR THE FUTURE

Unless there is a strong and effective political commitment to support the research base, Britain will fall even further behind in the quality of its research base and in its ability to be a figure of distinction among the top nations of Europe.
Sir Mark Richmond, Chairman of the Science and
Engineering Research Council, addressing a 1991
conference organised by the journal *Nature*.

The science base to which Sir Mark Richmond referred is beset by anxieties. There is a financial squeeze on government-funded research, to the extent that an increasing number of people are saying that our research effort is now over-extended. Meanwhile, university science departments are under threat of further contraction or closure. At the same time the funding of science in universities is facing two major changes. The first is the planned transfer of £154m, phased over three years (1992/3 to 1994/5), from the Universities Funding Council (UFC) to the Research Councils; the second is the proposed expansion of the entire higher education system. This latter initiative has two features: the conversion of the polytechnics to university status, which occurred during 1992, and the doubling of student numbers by the end of the decade. (The UFC and the Polytechnics and Colleges Funding Council are being replaced in April 1993. In their stead there are to be three Higher Education Funding Councils; one for England, Wales and Scotland. Northern Ireland higher education will come under the Northern Ireland office). In the school system there is a significant shortage of science and mathematics teachers, which is reflected in the reduced number of pupils taking science and mathematics at A level. In industry the science base is shrinking. Thus our total science base is under strain.

The core of these problems lies in the funding and management of research, because it is research activities which feed the science base in universities, in government scientific establishments and in industry. Research is also important as a barometer of institutional status and it is a means of attracting outside funds. In addition, research is seen as an activity which is inseparable from university teaching, though current financial pressures are starting to dilute what used to be an automatic link. On top of that, research performance is seen as the vehicle for career progression in the profession. Thus, research funding plays a central role in shaping the climate of science.

The tragedy at the moment is that while we still have much good science, we have no coherent national policy for funding and managing it. Three issues have traditionally dominated our policy needs: money, direction and management. How much can we afford, where should we place the funding and who should supervise it and according to what criteria? These three issues have been central to governments' management of science for decades, with the emphasis slightly shifting from time to time. In the 1960s and 1970s much of the debate was about management: what sort of bodies should have the responsibility for disbursing the funds? In the 1980s, the emphasis was placed on the direction of science funding, i.e. would it be commercially useful? In the 1990s the accent will be principally on the availability of funding, with research objectives and research management as subsidiary questions.

The present turmoil in science policy and funding might seem to be a relatively recent phenomenon but the contributory factors go back a long way. It is true that the problems are more intense than they were in the past. But the essence of our problems has been with us for a very long time.

Although British governments have been funding science to some degree since the nineteenth century, most of the framework for science funding dates from the period 1910–1920. The First World War showed up the country's lack of scientific knowledge in a number of industrial fields, prompting the establishment of the Department of Scientific and Industrial Research out of which grew, several decades later, the Science Research Council. This was followed by the Medical Research Council in 1920 and the Agricultural Research Council in 1931. Science was given a further boost after the Second World War with the setting up of atomic energy, aerospace and defence establishments.

By the 1950s, the funding of scientific research was coming from three main sources, central government departments, principally the Ministry of Defence, the Research Councils and the universities. An advisory body called the Advisory Council on Science Policy, composed largely of representatives of the Research Councils, existed, but its practical influence on research policy and funding was minimal. There was no overall national policy on science and no government minister responsible for science. There was, however, a government commitment to large projects in defence and aerospace which has remained to this day. British governments' relatively high spending on defence research and development – relative, that is, to their funding of civil research and development, and to other European countries' defence spending – has generally been the subject for criticism from scientists, and indeed economists. They both claim that this money has drawn funds and manpower away from more productive, commercially useful activities, and that any spin-offs into the commercial sector have been few and far between.

During the 1960s the priorities for science policy and its management began to be discussed more critically, forcing some major changes in government attitudes. Prime Minister Harold Macmillan announced the first Minister for Science but appointed a classics-trained barrister, Lord Hailsham to do the job. Despite his background, Hailsham is reckoned to have argued the scientists' case relatively well in Whitehall. Harold Wilson made a bolder attempt to lift the profile of science with the creation of the Ministry of Technology, under Tony Benn, and the appointment of the first Chief Scientific Advisor to the Cabinet, Sir Solly Zuckerman. Meanwhile, physics, astronomy and defence were taking a larger and larger amount of the total research budget and the economic benefits of research were not always apparent. A new advisory body was set up, the Council for Scientific Policy whose members were a mix of industrialists and scientists.

A radical shift in policies came in the 1970s as a result of recommendations made by Lord Rothschild. Government departments whose activities had a significant science component were to have a chief scientific officer who would act as advisor on scientific issues. The Council for Scientific Policy was reconstituted as the Advisory Board for the Research Councils with a much widened membership that included heads of research councils and government scientists.

A third recommendation was that the Research Councils' applied research should be governed by the 'customer contractor principle', i.e. that the customer (usually a government department) should dictate the terms of the work. Rothschild estimated that 25% of the Research Councils' work was in this applied area and that the funding for this work should come not from the Department of Education and Science (which funded the councils) but from the individual customer, i.e. government department. Therefore only 75% of their income would come from the DES with the rest coming from customers. This so-called 'Rothschild principle' was fiercely resisted, not only because it would undermine the independence of the councils, but also because it would force them to do some applied research in place of basic research. Eventually all the councils except the Natural Environmental Research Council successfully fought off the proposed cut in their DES funding. The NERC is still feeling the effects of the Rothschild principle in that it is more dependent on obtaining outside 'contract' work than are the other Research Councils.

An important innovation of the 1970s was the setting up of the Advisory Council for Applied Research and Development, which as with similar bodies, included both industrialists and scientists. ACARD issued a number of reports on major industries, but in practical terms its influence was small. Its report on Britain's medical equipment industry, for example, pointed to its significance in international terms and warned that it was slipping, but its recommendations were never adopted.

It was during the 1970s that the costs of science began to grow significantly faster than inflation. Margaret Sharp, a Senior Fellow at the Science Policy Research Unit, sees the mid-1970s as the turning point. Before then, she says, most good research proposals got funded, but after 1975 it became steadily harder to fund all the research that scientists wanted to do. It was in the 1970s, that Labour's Education Secretary Shirley Williams announced that 'the party is over for science'. With that pronouncement in mind it was therefore to be expected that many scientists would vote Conservative in 1979. But they were soon to be disappointed. The spending cuts that affected almost every Whitehall department in 1980 and 1981 had a dire effect on the sciences, especially engineering. Some extra money was set aside during the first half of the 1980s to create some new academic posts in science, colloquially called New Blood appointments, but the overall squeeze on science continued, and this was pointed out in a House of Lords Report on Science and Technology in 1982. By this time the Alvey initiative – a five-year information technology research programme involving universities and industry – was in full swing. The results from Alvey had a mixed reception; the commercial applications that accrued were thought to be disappointing, but this criticism missed the point that Alvey had been intended to foster collaborative research at the pre-competitive stage.

In the middle of the 1980s Prime Minister Margaret Thatcher decided to take special responsibility for science, by personally chairing a number of scientific committees. The accent now was on value for money and market-driven science, with universities and research agencies encouraged, or compelled out of financial necessity, to go out and bid for contract research of an applied nature. This philosophy remains a pervasive strand of current science policy. In the last few years, there have been several initiatives such as Inter-Disciplinary Research Centres (IRCs), LINK and ESPRIT which are all intended to focus on particular strategic areas of industrial or scientific importance, and which are designed to promote technology transfer.

In the latter part of the 1980s and early 1990s discussion moved back to the structure of funding. There were proposals to merge two of the Research Councils – the Agricultural and Food Research Council and the National Environmental Research Council – and to create a Biological Sciences Council. These concepts did not find much support. The creation of the IRCs was one structural innovation that did get off the ground. Another was the proposal to switch some funds from the UFC to the Research Councils. (The reasons for this will be discussed shortly.) Finally, there were some policy-makers, like Sir Mark Richmond, who called for a new integrated research funding structure.

In the aftermath of the 1992 Election, an Office of Science and Technology (OST) was created to be overseen by a Minister for Science. This effectively meant that science policy would be moved out of the Department of Education. At the same time, the new Minister for Science, William Waldegrave, announced that he would be launching a major review of British science policy.

A review of science policy needs to address seven issues:

1. The future of the dual support system for higher education funding.
2. The need for more money to support teaching and research.
3. The trend towards more selectivity in funding sectors in science and in funding science centres.
4. The role of 'strategic' research.
5. The proposed creation of a super-league of research centres.
6. The encouragement of research in small organisations and small universities.
7. More collaboration across disciplines, between research centres and across national boundaries.

1. WHITHER THE DUAL-SUPPORT SYSTEM?

The existing system of funding university research from universities' own income and from Research Council grants has broken down. In theory, in 1990/91 the Research Councils provided £821m for research, most of which went to universities, and the Universities Funding Council (UFC) provided a further £919m for research. In practice, a large part of the UFC money did not go towards university-generated research. It went to support research that was funded by the Research Councils.

The Association of University Teachers says that during 1988/89 the universities were diverting about a third of their own research funds to subsidising externally-funded research contracts, mostly Research Council projects. In money terms this switch of resources represented about £200m out of the total university research budget. What this has meant is that individual university departments have not had the money to finance their own research. One difference between university-funded research and that funded by the Research Councils is that the former has traditionally been an important source of support for young postgraduates who have yet to make a reputation, while the monies from the Research Councils (which tend to have fairly strict criteria attached to them) have gone primarily to those who have some track record. In these circumstances there have been far fewer opportunities for bright postgraduates to be given their own research programmes.

The current crisis in the dual-support system has a number of causes, which are partly accountancy-based and partly organisational. Firstly, the extent to which the universities indirectly contribute to costs arising from projects funded by the Research Councils has been underestimated for years. Externally-funded research requires the use of equipment and the employment of support staff. These indirect costs were never extrapolated by the universities' accounting systems, for the very good reason that it had not been necessary to have this detailed information in the past. Around 1991 the universities carried out an exercise to establish just how much they were spending to support external research; it turned out that the expense was much greater than had been thought – between £150m and £200m a year.

The second problem has been the lack of institutional communication between the UFC and the body responsible for the Research Councils, the Advisory Board for the Research Councils (ABRC). This appears to have been caused principally by the UFC's managing board, especially some of its past chairmen, who wanted to maintain a distance between the UFC and the ABRC. The UFC understandably felt that if it got too close to the ABRC, and revealed how it was spending its money, then this might compromise its independence. In fact, the UFC's stance only damaged itself, since the ABRC would not accept – until recently – that the universities were subsidising the Research Councils' work on a significant scale. With a new UFC chairman relationships have improved, but even now communication apparently leaves a great deal to be desired.

As things stand, the universities feel that they have been squeezed twice, firstly by the overall shortfall in university funding, and secondly by the need to support from their own resources some of their staff who are working on Research Council-supported projects. However, the Research Councils are not happy either because they still feel that their projects are not being supported either with the consistency or with the funds from the universities that they believe they deserve.

The government's response has been to announce the transfer of money from the UFC to the Research Councils in order to cover the overheads of research. This is being phased over three years, so that by 1994/95 the Research Councils will be receiving an extra £154m each year. The government claims that the total amount of funding for research will not fall – some of it will just be coming from a different source. Scientists are alarmed. Reducing the amount of funds that go to the universities for the purposes of research will make a big difference. They say that it will further weaken the dual-support system and that it will affect the kind of research that is done in universities.

The critical importance of university-funded research is explained by Professor Jean-Patrick Connerade of Imperial College:

> You can't equate research funds that come from the Research Councils with those that come from your university. Research grants from the Research Councils are won in competition with many other applicants and are awarded on the basis of fairly detailed proposals, which have to be adhered to and which require an annual report from the person in receipt of the grant. The university-funded research is undertaken at the discretion of the department head who is able to decide for him or herself what would be a worthwhile research project. It is the discretionary power and the flexibility to start or end a research project which is important and which is valued by academic scientists.

Connerade's concerns are mirrored by Professor Roger Cashmore at Oxford who warns of the uncertainty that lies ahead. He warns:

> The whole research support system during the change-over period is going to be confused. It will be administratively chaotic. People will not be clear what the rules will be for research grants and a lot of good research work will be up in the air. The prospect is appalling.

On the face of it, Cashmore's complaint sounds alarmist, but other scientists agree with him. The reason is that if the universities have less money, they will not have the funds to employ the technical support staff. Under the proposed system, most technicians will be paid by the Research Councils according to the research that they are supporting, which will mean that they will derive their income from various Research Council projects. It will thus prove a most complicated method of payment, and it will also give technicians even less job security than they now have. The cumbersome nature of this planned arrangement is further criticised by Professor Brian Heaton at Liverpool, who gives a specific example:

> We have a technician for our NMR machine for one hour a week, but we cannot aggregate that hour's work very easily to justify a separate claim for technical support for a particular project.

A deeper fear of scientists is that this £154m switch of resources might be the thin edge of the wedge, that all the research money that is now allocated to the universities might be in the gift of the Research Councils. Thus the Research Councils would become the single major source of government-funded research – there are other government departments who also fund some research, but of a very specific kind. The universities departments would have no money of their own for research.

Sir David Phillips, Chairman of the Advisory Board for the Research Councils, says categorically that there is no intention to remove research monies from the UFC or the new HEFCs. Indeed, he has publicly and privately affirmed his view that universities need to have their own source of research funds. He says:

> I believe that universities should continue to have unhypothecated grants to support research, because I believe that research leaders in universities who can see up-and-coming researchers are in a better position to identify and support such people than is some central committee. So I'm in favour of local patrons; I've benefited from that myself as a young scientist.

However, Sir David points out that the Treasury regards the university-funded research as 'a black hole', because universities do not have to account for how they spend this money and he therefore adds:

> We as a country have to be sure that the universities' money scheduled for research is actually going into research and in broad terms we need to know how it is being spent.

This view alarms the scientific community. Sir Michael Atiyah:

> To regard the university-funded research as a black hole is a wholly misconceived way of looking at universities. To get universities to say what they spend their research money on and how they identify research would be totally wrong; it would

mean treating research and teaching in a detailed way which just doesn't make sense. It certainly isn't a black hole; it just seems that way because people can't account for it. A lot of the money goes to supporting the salaries of a large number of staff engaged in research. That money is the basic underpinning that enables university staff to think, and if you take that away you totally alter the character of appointments in universities, and if you do that you will run down the quality of people in the higher education sector. Why should people go into the sector if they're not allowed to think, if their job is just to turn out graduates in a sausage machine? What attracts most people into academia is to be able to pursue ideas; if that's no longer part of their duty and if there is no time for that, why should people go into higher education? If the intellectual attractiveness is not there then you won't attract the best people, and that's what has happened in the schools for that reason.

For the moment the dual-support system remains in place and the universities have the freedom to spend money on research as they think best. But its continuity is not cast in stone. In the words of Sir David Phillips, its continuity 'depends on how transparently the universities use that money'. He goes on: 'The universities need to demonstrate that they are spending that money on research and spending it effectively, and if they don't there will be discussion of it being taken away from them.'

One of the most outspoken supporters of the current dual-support system is Sir Mark Richmond. He has argued that to dismantle it would be disastrous. In his address to the *Nature* conference he explained:

> High-class scientific research of the type needed to maintain a robust base for the present and the future must be located in our higher education institutions. Only then will be there be the necessary close coupling between research and the flow of young men and women capable of high-class research. A robust and well resourced dual-support system is a *sine qua non* for the provision of opportunity to our young researchers; and impaired opportunity is one of the most baleful influences on scientific research in our universities now. . .
>
> Multiplicity of choices is vital: no single panel of assessors is gifted enough always to make the right decisions on what to support.

We need to find a way of preserving the dual-support system and financing it adequately.

2. HOW MUCH MORE MONEY?

British science requires an infusion of extra money at three points: in research, in schools and in industry.

When scientists say they want more money, they usually mean that they need more money for research – in the universities and in the research institutes. The consensus among scientists is that about £300m/£400m extra a year would plug the shortfall in basic science funding, with about half the extra funds going to the

Research Councils and half going to the new Higher Education Funding Councils. In effect this would mean raising government support for what is called 'the science base' to 0.4% of GDP; this is what it broadly was in 1981/82, but it currently stands at about 0.28% of GDP. Scientists also advocate returning government support for civil research and development to at least 0.72% of GDP which is where it stood in 1981. In 1990 it amounted to 0.51% of GDP. Indeed British government support for civil research and development as a percentage of GPD is the lowest among the major European countries. Germany spent nearly 1% and France 0.85% of GDP on civil R&D in 1990. Our GDP itself is already lower than that of Germany, France or Italy; so it is clear that these countries are spending in real terms far more than we are on science in the civil sector.

In July 1992, the Government's own advisor, ACOST, went some way to recognising these arguments for more money, proposing that funding for the science base be increased by £762m, of which £460m should be for new equipment to support research 'at world-class level'. Putting more money into new equipment would give British science help where it is really needed. Sir Peter Medawar once made the point that while having good equipment does not buy ideas, the lack of equipment inhibits research that can only be evaluated by first-rate equipment.

Science also needs some financial support outside the higher education system. We need to put some extra money into the teaching of science and mathematics in schools. Two initiatives are required: one is to increase teachers' pay in order to attract science graduates into teaching, while the other is to set up a teacher training programme for older graduates, who may, for example, have lost their jobs in industry as a result of the recession or face redundancy in a defence-related establishment as a result of the reduction in our defence spending. Such scientists represent a potentially huge resource for the teaching sector, especially in physics and mathematics.

In industry, science and its exploitation might be enhanced if there were some tax concessions for research and development spending. Colin Humphreys, Professor of Materials Science at Cambridge, points to Japan, where companies are able to write off all their research and development expenditure for tax purposes. This is inevitably a controversial suggestion. It does not command the support of some eminent scientists-turned-businessmen such as Sir Robin Nicholson, who believes that tax benefits for industrial research and development would not work. But in a financial climate that rewards short-term thinking we do need some mechanism that would tilt the balance in favour of R&D spending in those circumstances where the boardroom arguments are finely drawn between investing and not investing. As a number of business people have pointed out, the single biggest inhibitor to company R&D spending is the high cost of capital. Bringing down interest rates is not an easy solution for it is restricted by domestic as well as international conditions. Changing the fiscal climate would be easier.

3. THE DRIFT TOWARDS GREATER SELECTIVITY

Greater selectivity in funding science seems inevitable. Indeed the process has already started. As Sir Mark Richmond has pointed out, 54% of the grants awarded by the SERC now go to eleven universities. These are, in alphabetical order: Birmingham, Cambridge, Edinburgh, Glasgow, Imperial College, London, Leeds, Liverpool, Manchester, Oxford, Southampton and University College, London.

However, the selectivity we have seen so far will probably have to be taken much further and it will also have to be more sophisticated, distinguishing more between departments and individual research centres than between universities.

There are three aspects to selectivity in science funding. There is the need to limit the number of centres which will receive the funds; this is coming to be recognised and despite its controversial implications it does seem the way of the future. There is then the separate, even more contentious, question of whether we can continue to fund every branch of science, i.e. should we start to focus on fewer areas of science? Can we continue to fund all aspects of astronomy and physics, which are both relatively expensive areas? Questions in this area have already been articulated in the scientific community and are often described in terms of 'big science' (i.e. expensive science) versus 'small science'. The third aspect of selectivity, which is the most difficult of all, involves the formal categorisation of universities into three groups: those that teach and research, those that mainly teach and do a little research, and those that only teach. The suggestion that universities should be classified in this way for the purposes of funding was made by the Advisory Board for the Research Councils in 1987. (See the ABRC's report *A Strategy for the Science Base.*)

These three modes of selectivity have all, to an extent, already been put into effect. Research funding is being concentrated in fewer centres; some subjects are being funded much more than others; and some universities are already being treated, in effect, largely as research centres or as institutions primarily concerned with teaching. But over the next three years the process may accelerate. The whole future of certain scientific subjects may be called into question and some universities may be removed from the research community altogether.

At the heart of the issue of selectivity lie questions about the future of certain sciences in a global context – about what discoveries may reasonably be anticipated in a particular field – as well as questions about the social and industrial applications that are foreseeable in certain scientific areas. Some of the greatest advances are being made in the so-called small sciences such as biology, biochemistry and genetics. The strides that have been made in these disciplines are opening up substantial opportunities for developments in medicine, surgery, agriculture and environmental protection. Some very difficult choices may well have to be made if the science budget does not receive the increase that it needs.

The most difficult of all may well be deciding which universities or which departments of universities should remain in the research sector. However, a number of scientists, such as John Mulvey, Secretary of Save British Science, argue that this is the wrong way of approaching the focusing of funds. In Mulvey's view the problem is 'not where should the money go, but to whom it should be given. Money should go to people, not departments.' To an extent this is how the MRC has distributed its funds; it has awarded grants to individuals and their teams which often cease when those people leave the institution where they have been working, but in practice the MRC still favours a number of institutions more than others. It seems likely that further selectivity will be made more on the basis of departments than that of individuals. If that does become the norm, then some mechanism will have to be found whereby very able scientists can transfer to the favoured departments. That at any rate, is the view of Sir Mark Richmond, though it is an idea that has yet to be fleshed out.

The third way in which selectivity is being increased is by the separation of higher education centres into those that mainly teach, those that research and those that combine the two in equal measure. The need to reasses the link between research and teaching has been given a new impetus by the decision to allow the polytechnics to acquire university status. Up till 1992 all universities got some money for research, no matter what their research standing. This was because the criteria for research funding was partly based on student numbers and partly based on research excellence. If this twin criteria had remained in place then it would have meant that every polytechnic that became a university would have the automatic right to research funding simply on the grounds of student numbers. The result would have meant that an already over-stretched budget would have been strained beyond its limits and money would have been taken from leading university science centres to fund brand new research departments in those institutions that had recently been made universities. So the rules had to be changed.

Under the new rules governing research funding the numbers of students a university has will not be a criterion for the award of money to do research. However, this new ruling will, in effect, protect some universities (those that have never relied on student numbers as a criterion for receiving research money) but leave others (those which have lots of students but which have only an average reputation in research) highly vulnerable. As Sir David Phillips bluntly explains: 'Some universities which have received research grants mainly because of student numbers are in for a shock.'

Even within the old university system (i.e. the institutions that had been funded by the UFC), there have been a number of science departments where there has been no research. But what alarms scientists now is that there are departments that have successfully combined research with teaching which will in practice find their research funds removed.

Yet there are a number of arguments to suggest that the historical links between research and teaching could very usefully be loosened. Firstly, the best researchers are not necessarily the best teachers. Academics who spend a large part of their time researching into a specific area within their chosen field of science do not always find it easy to move away from their area intellectually in order to take a broader view of their subject as a whole, which is what they need to do when they teach. A recent Cambridge graduate in zoology explains:

> Some of our lecturers were so immersed in their own narrow research that they were unable to look at the subject broadly; the discreteness of their research was reflected in the discreteness of their lecture presentations to the extent that there was no structure to their arguments.

Conversely some of the best teachers of science have done very little if any research. Another recent Cambridge graduate (in computer science) says:

> When you looked at around at the quality of teaching, you discovered that some of the most able lecturers had published very few research papers. They ranked low in the hierarchy but their contribution to science had been significant because it was often these good teachers who had inspired former pupils to take up science as a career, some of whom have become very distinguished.

Secondly, there are the practical arguments against linking research and teaching too closely. If given the choice, quite a number of researchers would not do any teaching at all; but they are obliged to do so by their contracts of employment. A number of scientists have told the author that they do not really like teaching unless it relates to their own specialisms. In addition many academics do not, realistically, have the time to teach while still fulfilling all their other functions – which often not only include research but also questing for funds and sitting on committees of various kinds.

Thirdly, a number of people argue that research should be a full-time activity without the distractions of teaching. Roy Pike, Professor of Electrical Engineering at Kings College, London, is a proponent of this point of view:

> I spent a good many years at the Royal Signals Research Establishment (RSRE), funded by the Ministry of Defence. RSRE has a record of turning out some excellent research. I believe that the quality of our research was immensely enhanced by the fact that we could concentrate single-mindedly on research projects to the exclusion of everything else.

However, these are not popular views among academic scientists. They feel that, if teaching is divorced from research, then the quality of teaching will suffer. If research is concentrated among a minority of universities, they say, with the majority acting principally as centres of teaching, then the overall standard of British science degrees will fall. In addition, scientists say that a higher education centre that only teaches is not really a university in the classical definition. As John Mulvey puts it, 'The basis of a university is that it is a place where knowledge is being advanced, even if only one or two groups are doing research.'

Yet underlying the pleas from academics that research opportunities should go hand in hand with teaching is the fact that research has a much higher status than teaching. Today, promotion is largely won on the basis of research achievements. Yet surely other criteria ought to be important as well, such as excellence in teaching – which is much rarer than excellence in research. The difficulty here is that it has been much easier to measure the output of research than it has been to measure the impact of teaching. There is surely a need for some mechanism by which those who communicate science well – not only to students but to the wider world – are amply rewarded and given the status that they deserve. One of the problems of science today – and it has become evident in other countries as well – is that it has become inward looking. Scientists need to be encouraged to communicate their subject. At present there is little inducement for them to come out into the open and talk about their work. Indeed some scientists – especially those who have been successful in the media – claim that there is a positive disincentive from within the profession to the wider discussion of its work. It really is time that the profession stopped regarding scientists who make teaching and writing their main activities as second rate.

4. IS THERE A NEED TO FOCUS ON STRATEGIC AREAS?

There are those who argue, in pressing for greater selectivity, that what we should be doing is concentrating our scientific efforts in areas of strategic economic or national importance. One such is John Fisher, Technical Director at PA Consulting Group, who puts the case like this:

> If you are going to be great as a country or as a business you have to focus, you can't do everything. Why can't we say, we are going to make the UK the number one in a particular industry or at least among the top players and then support all the research on which that industry relies and make our academics in that field the best in the world. Take medical instruments. The UK is doing appallingly in this field in general, but it is a major growth area and it is one in which we already have an awful lot of the base technologies. Why don't we set out to be the number one supplier in medical instruments? With the best science and technology base, the best research programmes, the best clinicians, and out of that we will build the best instruments companies in the world.

The concept of backing strategically important science gained a lot of ground in the late 1970s and early 1980s, and it inspired the five-year Alvey programme. Part of the rationale behind Alvey was that it would strengthen our 'enabling technologies', i.e. those essential technologies that enable an industry to compete in international markets. Enabling technologies in Alvey's sphere of information technology included silicon chips, software, computer-aided-design and robotics. Alvey has been succeeded by LINK, ESPRIT and a further IT initiative. Alvey also led to the creation of IRCs which bring together scientists from complementary backgrounds for specific research projects and which at the same time promote collaborative research with industry. Yet many of these projects have been insufficiently funded. An example is the semiconductor (silicon chip) IRC based at University College, London, which was seeking £3m at the beginning of 1992 in order to continue its work. There is a lot of strategic science being done, but it is skimpily financed.

Although there were some initial problems in the management of some of the IRCs, they appear to be working well. They are engaged on important research and they are acting as vehicles for transferring technology from the academic world into the business community. Yet the future of the IRCs is now under a cloud, because they are seen to be tying up money over a relatively long-term period, i.e. three to four years. Research Councils have always funded some long-term projects, but the requirement to fund the IRCs means that a bigger proportion of their money than they would like is now represented by long-term research. So there is less opportunity to support new projects. Therefore, there is the temptation to wind down the IRCs in order to increase the money available for shorter-term research. This surely does not make sense. If you want science to give you economic value for money, then the IRCs are clear candidates for continuing support. If this means that we have to spend more money on science as a whole, then we ought to be prepared to do that.

5. SHOULD WE CREATE A SUPER-LEAGUE OF RESEARCH CENTRES?

The rationalisation of our scientific research centres over the next three years resulting in the creation of a super-league seems inevitable. Private research institutes, government centres and industrial research establishments will all be affected by this trend. In particular it will affect the universities.

This scenario has its advocates and its critics. Among the supporters of the development is Derek Roberts, Provost of University College, London. He says:

> We have to be realistic. There are only two possibilities; either we inject a vast increase of funds into research or we become much more selective. It will undoubtedly be elitist, but if we go down the other route and say that every former polytechnic and university has to be considered equal then no one will be able to do anything excellent.

University College is already among the favoured – it has more inter-disciplinary research centres, for example, than any other university.

Those who are concerned at the prospect of fewer university research departments come from a wide spectrum of institutions not limited to those which could expect to suffer. At Oxford, for example, Professor Colin Blakemore worries that concentration of research will 'ossify the whole system'. He says:

> There has always been a hierarchy in research but at the same time the highest reaches of academe have been fed by new blood from the less well-endowed centres.

If we remove a whole swathe of research departments from the higher education sector then we will lose the possibility of that infusion. You need to have a circulation of people to keep the community of research healthy. On top of that, the elite centres could become complacent in the belief that their research funding is assured.

It is a point reinforced by Professor Denis Noble, also from Oxford:

I think that if we set things up in such a way that about ten universities will be assured of funds, then in ten years' time we'll reap the effects. Those universities will still do good work but I fear they will become complacent because they won't be challenged. There has to be the process of renewal.

In support of these arguments a number of academics refer to examples of well-funded departments in the UK that are to some extent resting on their laurels. These are institutions that have gained a reputation in the past, that have come to expect a high level of funding as of right, that have become a bit inbred and which as a result have seen their research success slipping in recent years.

Some scientists also say that studies in the USA and Britain comparing the output of large-scale research centres with that of small ones have found little evidence to support the view that there are benefits from economies of scale. But the evidence is mixed. One US survey into biomedical research (McAlister & Narin, 1983) showed that while the size of research centre was not an indicator of publication output, the larger institutions did have higher citation rates, i.e. that their research papers were cited by other scientists more frequently. In the scientific world, a higher citation rate is taken to mean that a piece of research is better regarded than others.

Some scientists have suggested that we will have to work out a compromise solution. They say that the creation of super-league research centres crammed with the best equipment is inevitable, but at the same time they are suggesting that these centres should be open to scientists from other universities. One such is Dr Brian Richards, Chairman of British Biotechnology and Chairman of the Biotechnology Joint Advisory Group. He says:

We ought to move towards having a series of research hotels, which would be laboratories having a core of top quality people, in which other scientists can work for a period of time. In the life sciences, for example, (biology/biochemistry/molecular biology) there are about seven sites in the UK which all have the best equipment. People should be able to work in them to do their own research, working intensively for a term perhaps, and then to go back and teach in the other three terms.

He adds:

Of course it would be domestically upsetting for those coming into the research hotels, but at least it would mean that they wouldn't be denied access to high quality research equipment. It would give everyone an opportunity to contribute to the nation's research and in an atmosphere of very high standards.

In physics, the sharing of facilities is already happening to a limited extent, he says, but it could be taken much further.

6. DEVELOPING SMALL RESEARCH ORGANISATIONS

Alongside the super-league of research centres there will be a need to support a number of smaller centres of research excellence. It is undeniably desirable to maintain some diversity in the academic system – and even in science big is not always beautiful.

Max Perutz points out that a lot of good research has been done by quite small teams. He cites the invention of monoclonal antibodies, which was the work of two people, and high temperature superconductivity which was discovered by two researchers in Zurich. Perutz also says that much good research work is done in departments in small universities. Indeed he argues that some of the most original ideas have sprung from small universities. Examples he gives include George Gray's work on liquid crystals at Hull, mentioned earlier, vibrational spectro-scopy of surfaces which was due to Sheppard at East Anglia, magic angle spinning in magnetic resonance imaging from Andrew at Bangor, crystal structures research by Woolfson at York, and Sir John Meurig Thomas's work on solid state catalysts and crystalline solids at Aberystwyth. Another example is the work on DNA fingerprinting by Alex Jeffries at Leicester.

Arnold Wolfendale, the Astronomer Royal, takes a similar view, saying that 'most research is undertaken by small groups, sometimes even by individuals', and that small groups can flourish equally well in small and in big universities. He also argues that the research information that is acquired on the expensive projects in the big universities will yield the greatest benefits if disseminated to as wide a group of people as possible, and these should include small research teams.

Wolfendale also believes that much of science may have to go the way of astronomy, where the days when each university could expect to have its own telescope are long since past. Astronomy in Britain is carried out, he says, by sharing resources on a national and international scale.

Several scientists have emphasised that it would be foolish to assume that all the discoveries of the future are going to come out of big, expensive laboratories. Two scientists in particular were anxious to counter that popular assumption. They are Sir Michael Atiyah and Don Braben, who set up BP Venture Research. Sir Michael argues:

> If you just go on supporting the advanced high-tech science you are maybe narrowing your focus into areas that are possibly past their prime so that you get diminishing returns. You ought to leave some money for those things that are unfashionable, for the sciences of the next generation which often do not require expensive equipment.

Don Braben, who backed a number of unusual scientists as manager of BP Venture Research, takes an even stronger line:

If you focus more and more on goal posts, especially goal posts that scientists in other countries are aiming at, then you force scientists to have a competitive edge. If you are all going in the same direction, then that is bound to put pressure on the acquisition of expensive equipment. The more competitive the area of science, the more expensive it becomes. But if you start in a new field, where nobody else is going, you don't need expensive equipment.

Braben cites his own experience at BP Venture Research. There, his budget for funding scientists – outside BP – who wanted to research unusual ideas ranged between £2m and £3m a year. With this relatively small sum he was able to fund more than thirty scientists a year. Many of them are now working on industrial applications of their original ideas.

7. MOVING TOWARDS COLLABORATION

Science thrives on collaboration as well as on competition. In the future we shall have to have more collaboration between different research centres within the UK as well as across national boundaries. Sir Arnold Wolfendale says that there is already some very effective collaboration between different university departments in the UK. Internationally collaboration is also growing. Britain already belongs to three international projects in physics. At the same time individual departments are working with partners overseas, especially in medicine and physics.

This international collaboration is bound to increase, says Sir Michael Atiyah, but in order for it to function well, he says, it needs new structures.

> There are a lot of structural differences between Britain and the rest of Europe in the way science is funded – a lack of harmony. We want to see some convergence in funding structures if we are going to work with other countries.

He and other scientists cite the example of France and Germany which have two basic channels of funding – one to fund internal fundamental research and another one to fund big international projects – and these monies come out of separate ministries. In France for example, the big international projects are funded by their Foreign Office.

Although more cooperation across Europe may be desirable, scientists complain that the grant application procedure for European Commission funds is even more complicated than in Britain. They all say that it is unbelievably bureaucratic.

One stipulation is that the recipient of an EC research grant can only employ EC nationals on a project. But, say scientists, this is not always feasible. One chemistry professor has a project in mind which would require an assistant researcher but he says that he cannot find a suitable person from within the EC to fill the post. Another requirement is that EC projects should be collaborative, i.e. they should involve two, or preferably more, research centres spread across several countries. According to one scientist, the Brussels civil servants appraising these applications like to fund 'laboratories without walls', large networks of scientists spanning five or six countries. Trying to conduct research in those circumstances will become counter-productive, say the scientists. The research becomes committee-driven, rather than ideas-driven. Yet many scientists have nevertheless applied for EC research grants, for the simple reason that the funds available are huge.

Some scientists are also concerned that British participation in EC research funding is siphoning money away from the British science budget. They already see that the funds needed to contribute to international facilities in physics and astronomy are leaving a smaller and smaller proportion available for genuinely British science. Other countries like Germany and France have much larger science budgets, so their international subscriptions and contributions to EC research budgets are proportionately less onerous. The trend towards what they see as the centralisation of funds outside Britain worries these scientists a great deal.

British science needs more money, but it also requires some new institutional structures in order to grapple more successfully with the issues outlined here. At the moment, change is happening in a piecemeal, almost haphazard, fashion with departments closing down by default and research projects left high and dry. At the same time, the scientific profession has fallen even more out of favour and seems rudderless. We need some dramatic change to rectify the state of British science. The next chapter outlines an initiative that could do much to transform science in Britain – if it were to be implemented boldy but sensitively.

— *14* —
ESTABLISHING A SCIENCE DIRECTORATE

If we are to strengthen our efforts in science, to use our intellectual resources more efficiently and to maximise the output of science, then we need to do more than just raise the level of science funding. We need a new and radical framework for British science. What is required is a new institutional and funding mechanism to bring coherence, vision and leadership to our national science policy and its management.

There are two critical requirements. The first is that science needs to be represented at national level in a high-profile way. The appointment of a Minister for Science is welcome, but it is not sufficient on its own. This post is a political one, and the Office of Science and Technology has yet to demonstrate its influence. Science needs to be represented by someone outside politics who is nevertheless funded from the public purse, who has achieved eminence in science or industry and who commands the respect of politicians, scientists and industrialists.

The second requirement is to bring together the various bodies that have an interest in the funding and exploitation of science, as well as those who currently have the responsibility for disbursing science funds. The current institutional framework is too fractured. It is a mess. We have various advisory bodies such as the Advisory Council for Science and Technology, the Council for Innovation and the Centre for the Exploitation of Science and Technology as well as the Advisory Board for the Research Councils, all of which have overlapping remits and, effectively, little power.

We have funding institutions such as the Research Councils and the Universities Funding Council (now being transformed into the Higher Education Funding Councils) which operate within a cloud of secrecy, provoking dissatisfaction among scientists and among the users of science, and this dissatisfaction is deepened by the fact that, in practice, they are not sufficiently accountable to a higher body. In addition there is no national mechanism for deciding strategic priorities or for mediating in disputes affecting competing science interests.

The present institutional arrangements are extraordinarily Byzantine and they are currently working against the best interests of science, as Sir Mark Richmond points out:

The predicament of the SERC and all the Research Councils is that they are serving a number of masters. We are serving the needs of curiosity-driven research and we are also funding work that is supposed to be useful, which is why, for example, we put money into engineering. Then you have the Advisory Board for the Research Councils (ABRC) which owes allegiance to the Department of Education and Science [now being transferred to the Office of Science and Technology] while engineering comes under the Department of Industry. If you take the Medical Research Council they are linked closely to the Department of Health, but neither of these departments figure explicitly in the dicussions of the ABRC.

Then there is the Universities Funding Council. Until relatively recently there was a barrier between the UFC and the ABRC although we are both funding university research! Under the new management of the UFC there is much more communication between us, but the Chairman of the UFC is not a member of the ABRC. We've asked that he should be. How can we assess where money should go to in the universities if we don't know where the Universities Funding Council is putting its money? There is no interlocking of policy.

The current institutional set-up seriously interferes with a coherent policy for the exploitation of science, as Sir Mark also points out:

An important question at the moment is what is the role of the Research Councils and the science base in relation to industry and commerce? Are we acting as a research reservoir for high-technology companies, are we supposed to be doing the strategic research that would be exploited by high-tech companies and is there any purposive coupling of us to those companies? At the moment there isn't any, or at least very little.

He continues:

We see nothing of what the Centre for the Exploitation of Science does or what the Advisory Council for Science and Technology does before it reaches the public.

The proposal to set up Faraday centres on the lines of the German Fraunhofer Institutes, which was developed by CEST, ACOST and the Department of Industry, came completely out of the blue, but we'd been working on a similar idea – the Parnaby centres – for a year. There's a missing cohesion.

What is even more bizarre is that one of ACOST's members, Sir David Phillips, Chairman of the ABRC, on which Sir Mark Richmond sits, did not communicate this development to the ABRC. Sir David insists that he cannot disclose the discussions of ACOST to the members of the ABRC, because ACOST is constitutionally set up as an advisory body to the Cabinet and therefore its deliberations are supposed to be secret. Several members of the ABRC have pressed Sir David to tell them something of the collective thinking of ACOST, because it would assist the ABRC in formulating science policy. But he says he is not allowed to. This secrecy between government bodies that are all trying to shape science policy has become extremely damaging as the farce of two organisations separately laying plans for near-identical institutions (Faraday and Parnaby) starkly demonstrates.

To get round these problems, Sir Mark Richmond has proposed setting up an Advisory Board for Research (ABR), which would bring together people from the Research Councils, the UFC and the major government departments. It is not intended, he says, to establish a big brother organisation but to bring in 'a sense of coherence'. He explains: 'I see the ABR as a communications centre as to what people are doing in different areas to achieve coherence.' Another proposal – from the leading science journal *Nature* – is for the scrapping of the Advisory Board for the Research Councils and its replacement by a publicly accountable Science Advisory Council, on the lines of a similar body in the USA. These proposals, especially that of Sir Mark Richmond, have much merit in them.

But in some respects the proposals are too ambitious, while in others they do not go nearly far enough. Sir Mark Richmond's Advisory Board for Research sounds similar to the Central Advisory Council, which was set up in 1967 under Sir Solly (now Lord) Zuckerman when Harold Wilson was prime minister. However, as Lord Zuckerman points out in his memoirs, this Council had a lifespan of only four years. Cabinet ministers, especially those whose departments had a direct interest in science – Defence, Education and Industry – complained of its interference. Part of its brief had been to comment on research and development policies in every government department. This caused a good deal of resentment especially at the Ministry of Defence. Sir Mark Richmond has not suggested that an Advisory Board for Research would overtly seek to influence the research policies of powerful government departments, but that would be the end result. If that happened it would meet with the same resistance that the Central Advisory Council did. In any case, the review of government departments' research is probably best undertaken by the Office of Science and Technology, in consultation with the Chief Scientific Advisor to the Cabinet, which is what has been proposed. However, something like the ABR, which sought to bring together the Research Councils and the Higher Education Funding Councils ought to be given serious consideration.

There remains one problem that always attaches itself to any semi-independent body that sits within the Whitehall machine: it lacks independence. It might be thought that a body located firmly outside Whitehall would lack the easy conduits of comunication to government departments. Not so. Contrary to what one might think, the proximity of such a body to other government departments does not automatically mean that it knows what is going on in those departments or, even more importantly, that it can influence them. Lord Zuckerman makes this very plain. The various scientific councils that he sat on only had influence to the extent that they were privy to what was going on 'behind the closed doors of government departments'. Often they were not. Therefore, it is far better for a new advisory and funding body to be rooted beyond the reach of the government machine.

What is needed is to rationalise some of our national science institutions and bring them together under a supervisory body with substantial funds and with real clout. This is not an argument for the over-centralisation of funding. It is an argument for achieving effectiveness, cooperation and the integration of policy-making for the benefit of science. These two requirements – the creation of a high-profile appointment representing science and the setting up of an institution responsible for a coherent science strategy – could be achieved by establishing a Science Directorate. It would be a statutorily established organisation, like the Arts Council or the British Broadcasting Corporation, but independent of politics. With a membership of no more than fifteen, about half of whom would be drawn from the scientific world, with the rest either representing industry and some other interest groups or acting as independents.

The Science Directorate would be funded on a three-to-five year basis, receiving its income from the Department of Education and Science or the Office of Science and Technology, and it would dispense funds to three bodies. Having an assured amount of funds over a 3–5 year period would give it the flexibility and the capacity to apportion grants on a longer-term basis than is currently possible.

The largest chunk of its funds – about £1.1bn – would go to what is now called the Advisory Board for the Research Councils, which would be re-named the Research Councils Directorate. The Research Councils Directorate would remain an independent body with stronger powers than it has today, but would be accountable for its decisions to the Science Directorate. Its chairman would automatically sit on the Science Directorate.

A slightly smaller sum – about £1bn – would go to the Higher Education Funding Councils and would be earmarked for university research. It would not be allowed to be used for day-to-day university spending and administration, nor would it be used to subsidise Research Council research. The chairman of the HEFC would also sit on the Science Directorate.

A third part of the Science Directorate's income would go to a new body which would be called the Applied Science and Technology Directorate (ASTD). This would be created by bringing together the Advisory Council for Science and Technology and the Centre for the Exploitation of Science and Technology. Ideally the ASTD would be funded half by the Science Directorate and half by manufacturing industry, the service sector, and enlightened financial institutions, and would have a total annual income in the order of £500m. The aims of the ASTD would be threefold: to fund strategic research (lying between basic and applied research), to fund particular pieces of applied research and to supervise current applied research programmes coming under the LINK and ESPRIT schemes. Today's LINK steering committee would be subsumed within the ASTD.

The £250m from business would be mainly derived from the top 1,000 companies and would represent less than 0.5% of their profits. With an annual income of £500m there would be the opportunity for creating Alvey-type projects.

These would be drawn up by the ASTD itself in collaboration with the Research Councils Directorate and the HEFCs, with the Science Directorate acting as a forum for some of the discussions. To enhance the effectiveness of the ASTD, not only would its chairman sit on the Science Directorate, but it would also draw some of its members from the Research Councils, the universities and the HEFCs.

Yet it would be a mistake if the ASTD were merely to be a club for big business and the big research institutes. There would need to be some representation from smaller research centres and from the smaller, more entrepreneurial businesses. The ASTD would be failing in its job if it did not help in the development of the smaller firms.

The Science Directorate would also have another function, which would be to promote science itself in the same way that the Arts Council encourages appreciation of the arts. It could, for example, give awards to young scientists in schools and in industry and to companies for excellence in science. It could arrange regional and national science festivals, drawing on contributions from museums, art galleries, libraries, schools, universities and business.

If these changes were enacted and if the new body had as high a profile as the Arts Council, not only could it help to give a sense of direction to British science but it could also – like the Arts Council – help to bring science itself firmly within British culture.

The chief executive of a Science Directorate or a Science Council or whatever one were to call such a body would be able to speak with independence and authority. Being in command of substantial funds, he or she would also be listened to. Lastly, they would have the means to implement policy.

It makes no sense to have a system where the best minds – the policy advisors and the scientists – have little or no power, while those with the power – the politicians, the civil servants and the funders – have an inadequate appreciation of the long-term economic and strategic issues involved. We are getting the worst of both worlds under the current structural arrangements.

There is also the further point that a body like the Science Directorate with a chairman appointed for a fixed term, say for five years, would lend some stability to science policy. The great Hurrah that we have a Minister for Science with a proper department at last, overlooks one fact. Cabinet ministers in Britain rarely stay in their posts longer than about two years. This will be true for the OST as it has been for every other ministry. Do we really want to continue to make major switches in science policy every three years, which is what happened in the 1980s; or do we want some more permanent framework that will outlive the oscillations of governments' ideologies?

SUGGESTED FURTHER READING

The following citations are a small sample of additional reading, many of which contain further source references.

BOOKS

Jacob Bronowski: *Science and Human Values* (Hutchinson, 1961).

Jacob Bronowski: *A Sense of the Future* (MIT Press, 1977).

Francis Crick: *What Mad Pursuit* (Penguin Books, 1990).

Peter Mathias (editor): *The Transformation of England – Essays in the Economic and Social History of England in the Eighteenth Century* (Methuen, 1979).

Sir Peter Medawar: *The Limits of Science* (Oxford University Press, 1985, and reprinted 1986, 1989).

Sir Peter Medawar: *Memoirs of a Thinking Radish* (Oxford University Press, 1986).

Keith Pavitt and Michael Worboys: *Science, Technology and the Modern Industrial State* (Butterworth, 1977).

Max Perutz: *Is Science Necessary? – Essays on Science and Scientists* (Oxford University Press, 1991).

Nathan Rosenberg: *Inside the Black Box* (Cambridge University Press, 1982).

C. P. Snow: *The Two Cultures* (Cambridge University Press, 1959).

Martin J. Weiner: *English Culture and the Decline of the Industrial Spirit, 1850–1980* (Cambridge University Press, 1981).

Tom Wilkie: *British Science and Politics since 1945* (Blackwell, 1991).

Lewis Wolpert and Alison Richards: *A Passion for Science* (Oxford University Press, 1988).

Solly Zuckerman: *Monkeys, Men and Missiles: His Second Volume of Autobiography 1946–1988* (Collins, 1988).

COMPENDIUM BOOKS ON SCIENCE AND THE TWENTIETH CENTURY

The following three books provide an interesting overview of the history of science, particularly in this century. The first two give a useful international perspective against which to assess the contributions of British Science.

The Nobel Century (Chapman, 1991).

Science: Invention and Discovery in the 20th Century (Harrap, 1990).

They Made our World: Five Centuries of Great Scientists and Inventors (Broadside Books Ltd, 1990).

ARTICLES, REPORTS, LECTURES

'Attitudes to R & D and the application of technology: Survey undertaken in France, Japan, The Netherlands, United Kingdom and West Germany, 1989' (PA Consulting

Group, London, 1989).

Colin Blakemore: 'Who Cares About Science?'. Address to the 1989 annual meeting of the British Association for the Advancement of Science, published in the journal *Science and Public Affairs*, Vol. 4, 1990.

Don Braben: 'Science and Economic Growth', *Interdisciplinary Science Reviews*, Vol. 14, No. 4, 1989.

Jean-Patrick Connerade: 'Public Perceptions of Science – Pursuit of Knowledge or Engine of Profit?'. Published jointly by the Science Policy Research Unit and Imperial College, London, 1988.

Ali El-Ghorr: 'In search of a structured career', *New Scientist*, 18 April 1992.

'Innovation, Investment and the Survival of the UK Economy'. Papers presented to a conference organised by PA Consulting Group, the Institution of Mechanical Engineers, the Fellowship of Engineering and British Aerospace, July 1989.

Peter Mathias: 'The Value of Useless Science', *Physics World*, January 1989.

Nature's 'Manifesto for British Science', *Nature*, 12 September 1991.

Denis Noble: 'Scientists have souls, too', *The Independent*, 24 December 1991, p. 15.

Keith Pavitt: 'What we know about the strategic management of technology', *California Management Review*, 32, 1990.

Sir David Phillips: 'Funding the UK science base – modes of support'. Lecture delivered at The Royal Society, 25 October 1988, and published in the journal *Science and Public Affairs*, 1989.

Sir Mark Richmond: 'A Future for British Science'. Speech delivered to a conference organised by *Nature* and published by *Nature*, 3 October 1991.

Sir Denis Rooke: 'Some Perspectives of a Simple Engineer'. Address to the 1991 annual meeting of the British Association for the Advancement of Science.

Nathan Rosenberg: 'Critical issues in science policy research'. A paper presented to a conference to mark the 25th anniversary of the Science Policy Research Unit, Sussex University, 1991, and published with other conference papers in Science and Public Policy, December 1991 (Beech Tree Publishing).

Margaret Sharp: 'Neglecting the Seedcorn – is Britain doing enough to nurture its young scientists?', *Science and Public Affairs*, Vol. 6, 1991.

GOVERNMENT PUBLICATIONS

Research in the United Kingdom, France and West Germany: a Comparison (SERC, 1990).

A Strategy for the Science Base, prepared by the Advisory Board for the Research Councils (HMSO, 1987). This was an important and controversial policy document, which set out the case for grading higher education institutions into 'R', 'T' and 'X' categories:

  Type R: offering undergraduate and postgraduate teaching and substantial research activity across the range of fields;
  Type T: offering undergraduate and M.Sc. teaching with associated scholarship and research activity but without advanced research facilities;

Type X: offering teaching across a broad range of fields and substantial research activity in particular fields, in some cases in collaboration with others.'

Annual Review of Government Funded R & D, 1982/1992 (HMSO).

ANNUAL REPORTS OF THE SERC, AFRC, MRC AND NERC

Innovation in Manufacturing Industry, House of Lords Select Committee on Science and Technology (HMSO, 1991).

Civil Research and Development, Report by the House of Lords Select Committee on Science and Technology (HMSO, January 1986). (See also the submissions and oral evidence – from universities, polytechnics, industry and others – given to the Committee).

Science and Technology: A Review of the Issues (ACOST 1992).

Evaluation of the Alvey Programme for Advanced Information Technology, Report by the Science Policy Research Unit, University of Sussex and the Programme of Policy Research in Engineering, Science and Technology, University of Manchester (HMSO, 1991).

Civil Exploitation of Defence Technology, Report to the Electronics EDC by Sir Ieuan Maddock, and observations by the Ministry of Defence (National Economic Development Office, 1983).

INDEX OF ACRONYMS

ACARD   Advisory Council on Applied Research and Development to the Cabinet Office.
ACOST   Advisory Council on Science and Technology to the Cabinet Office.
AFRC   Agricultural and Food Research Council.
CERN   European Laboratory for Particle Physics.
CNAA   Council for National Academic Awards.
FIEE   Fellow of the Institution of Electrical Engineers.
FRAS   Fellow of The Royal Astronomical Society.
FRCP   Fellow of The Royal Society of Physicians.
FRS   Fellow of The Royal Society.
FRSC   Fellow of The Royal Society of Chemistry.
FRSE   Fellow of The Royal Society of Edinburgh.
MIT   Massachusetts Institute of Technology.
MRC   Medical Research Council.
SERC   Science and Engineering Research Council.

BIOGRAPHICAL INDEX

The following biographies list principal academic or industrial appointments plus membership of various committees. They exclude publications. (Most interviewees have published extensively.) In most cases, more detailed information can be obtained from *Who's Who* or Debrett's *People of Today*.

ABEL, Edward, FRS, Professor of Inorganic Chemistry, Exeter University since 1972 (Head of Department, 1977–1988). Previously, Lecturer, then Reader in Chemistry, Bristol University, 1957–1971. Member of the University Grants Committee, 1986–1989; CNAA, Chairman of Physical Sciences Committee since 1987; Permanent Secretary, International Conference on Organometallic Chemistry, 1972–1989.

ALEKSANDER, Igor, FIEE, Professor of Electrical Engineering and Head of Department, Imperial College of Science and Technology, London University since 1988. Previously Professor and Head, Kobler Unit for Management Information Technology, 1984–1988, (Imperial College), Professor and Head of Electrical Engineering Department, Brunel University, 1974–1984.

ALFORD, Dr Neil, Research Manager, ICI Chemicals and Polymers Division, Runcorn.

ANDREW, Richard, Professor of Animal Behaviour, Sussex University since 1968, Chairman of the Biological Sciences Committee, SERC. Formerly Reader in Animal Behaviour, Sussex University (1965–1968), Assistant Professor of Zoology, Yale University.

ATIYAH, Sir Michael, FRS, FRSE, President of The Royal Society since 1990, Master of Trinity College, Cambridge University, since 1990, Director, Isaac Newton Institute for Mathematical Sciences since 1990; Royal Society Research Professor at the Mathematical Institute, and Professional Fellow, St. Catherine's College, (both Oxford University), 1973–1990; Professor of Mathematics, Institute for Advanced Study, Princeton University, 1969–1972; Member of SERC, 1984–1989.

BELL, Jocelyn (married name BELL BURNELL), FRAS, Professor of Physics, The Open University since 1991. Formerly at The Royal Observatory, Edinburgh. Other previous posts include: Research Fellowship, Southampton University, 1968–1973; Research Assistant Space Science Laboratory, University College, London University.

BIRCHALL, Derek, OBE, FRS, FRSC. Professor at Keele University since October 1992. Formerly ICI Senior Research Associate, Mond Division, since 1975. Visiting Professor, Surrey University, 1976–1988, MIT, 1984–1986, Durham University, 1987–1992.

BLAKEMORE, Colin, Waynflete Professor of Physiology, Oxford University, since 1979, Director of the McDonnell-Pew Centre for Cognitive Behaviour, Oxford, since 1989; Associate Director, MRC, Oxford Research Centre in Brain Behaviour since 1991; Royal Society Locke Research Fellow, 1976–79. Visiting Professor New York University, 1970, MIT, 1971.

BLUNDELL, Thomas, FRS, Deputy Chairman and Director General of the Agricultural and Food Research Council since 1991. Director of the International School of Crystallography, Council Member of the SERC, Member of MRC Aids Research Committee, Member of ACOST, Hon. Director of the Imperial Cancer Research Fund Unit of Structural Molecular Biology. Advisor to Abingworth Management Ltd (venture capital company).

BRABEN, Dr Donald, Chief Executive of Venture Research International. Formerly head of Venture Research, BP International. Previously advisor to the Bank of England and the Cabinet Office's Chief Scientific Advisor.

CASEY, Harry, FRSC, analytical and water chemist, Institute of Freshwater Ecology's River Laboratory, East Stoke, Wareham.

CASHMORE, Roger, Professor of Experimental Physics, Oxford University since 1991 (Lecturer, 1979–1990, Reader 1990–1991). Research Associate, Stanford Linear Accelerator Centre, California, 1969–1974. Various academic posts at Oxford University 1974–1979. Member of the Committees for CERN, SERC, and Deutsches Electronen-Synchrotron, Hamburg.

CONNERADE, Jean-Patrick, Professor of Atomic and Molecular Physics, Imperial College of Science and Technology, London University since 1985 (Reader 1973–1979, Lecturer 1980–1985). Since 1969, guest researcher at the Physikalisches Institut of Bonn University.

DAVIES, Paul, Professor of Mathematical Physics, Adelaide University. Formerly Professor of Physics, Newcastle University.

DAWES, Dr Bill, Lecturer at the Whittle Laboratory, Department of Engineering, Cambridge University. Previous positions were held in the Central Electricity Generating Board and in the Turbine Group as a Research Officer.

DEVONSHIRE, Dr Alan, biochemist and plant geneticist, Institute of Arable Crops, Rothamsted Experimental Station (Head of Insecticide Resistance Group), Harpenden. Special Professor, Department of Genetics, Nottingham University.

DOLPHIN, Dr Annette, Professor of Pharmacology, Royal Free Hospital School of Medicine, London. Lecturer, St George's Hospital Medical School, London, 1983–1990; Researcher at the National Institute for Medical Research, Mill Hill, 1980–1983; Researcher at Yale Medical School 1978–1980; Researcher at the College de France, Paris, 1977–1978.

DUFF, Dr Keith, Director of Research at English Nature (formerly part of the Nature Conservancy), Huntingdon.

DUFFETT-SMITH, Dr David, Lecturer in astronomy and researcher at the Cavendish Laboratory, Cambridge University.

DURANT, John, Professor of History and Public Understanding of Science, Imperial College of Science and Technology, London, since 1989; Head of Research and Information, Science Museum, since 1989.

EFSTATHIOU, George, Savilian Professor of Astronomy, Oxford University since 1988. Previously Senior Research Fellow, King's College, Cambridge, 1984–1988; SERC Research Assistant, Institute of Astronomy, Cambridge University 1980–1983.

EISNER, David, Professor of Physiology (specialising in veterinary medicine), Liverpool University. Former post-graduate student (heart arrhythmias) of Professor Denis Noble.

ELLINGTON, Dr Charles, Lecturer in Zoology, Cambridge University.

ELMS, Dr Robert, biologist, Furzebrook Research Station, (part of the Institute of Terrestrial Ecology).

FISHER, John, Technical Director of PA Consulting Group's Cambridge Laboratory. Winner of two Design Council awards and the Duke of Edinburgh's Designer's Prize. His work on a tonometer for Keeler Ltd, won for Keeler and PA a Queen's Award for Technological Achievement in 1989.

GOWENLOCK, Brian Glover, CBE, FRSE, FRSC, Professor of Chemistry, Heriot-Watt University 1966–1991; Member of UGC 1976–1985, Vice-Chairman of UGC 1983–85.

GRAY, George, FRS, FRSE, FRSC, Advanced Material Consultant, Merck Ltd. Staff of Hull University since 1946, Reader, 1964, Professor of Organic Chemistry 1978, Hon. Professor, 1990. Visiting Professor, Southampton University since 1990. Awarded Queen's Award for Technical Achievement, 1979.

GREGORY, Richard, CBE, FRSE. Professor of Neuropsychology and Director of Brain and Perception Laboratory, Bristol University, 1970–1988, now Emeritus. Formerly, Lecturer at the Department of Pyschology, Cambridge University, 1953–1967; Professor of Bionics, Department of Machine Intelligence and Perception, Edinburgh University, Chairman of Department 1968–1970. Founder and Chairman of the Trustees, The Exploratory Hands-on Science Centre.

HEATON, Brian, Grant Professor of Inorganic Chemistry and Head of Department, Liverpool University.

HERRIOT, Walter, Director, St John's Innovation Centre, Cambridge. Previously small firms advisor with accountants Coopers & Lybrand. (His comments in the book are based on earlier conversations with author and also on his contributions to a seminar on Innovation, December 1991, organised by the Wolfson Cambridge Industrial Unit, Cambridge.)

HEWISH, Antony, FRS, Professor of Radioastronomy, Cambridge University, 1971–1989, now Emeritus; Fellow of Churchill College, Cambridge, since 1962. Awarded Nobel Prize for Physics (jointly) in 1974.

HOOLEY, Richard, bio-chemist at the Long Ashton Experimental Station, Bristol.

HOOPER, Dr David, biologist, Head of Monks Wood Experimental Station, (part of the Institute of Terrestrial Ecology and the Institute of Freshwater Ecology).

HOPKINS, Harold, FRS, Emeritus Professor of Applied Optics, Reading University since 1984 (Professor, 1967–1984). Previously Research Fellow, then Reader in Optics, Imperial College of Science and Technology, London University, 1947–1967.

HORNE, Dr Nigel, Information Technology Partner KPMG Peat Marwick McLintock (accountants/management consultants). Member of the Cabinet's Advisory Council on Science and Technology (ACOST). Previously chairman of the DTI/SERC IT Advisory Board 1988–1991. Various GEC posts: Management Systems Manager, 1970–1972, Manufacturing General Manager, 1972–1975, Director and General Manager, Switching, 1976–1982. STC Director:Technical and Corporate Development, 1983–1990.

HOUZEGO, Peter, Principal Consultant, Physics Group, PA Consulting Group's Cambridge Laboratory. Formerly a senior manager in the electronics industry.

HUMPHREYS, Colin, Professor of Materials Science, Cambridge University since 1990. Previously Professor of Materials Engineering and Head of Department, Materials Science and Engineering, Liverpool University 1985–1989. Visiting Professor, Illinois University, 1982–1986. Chairman of Materials Commission and Engineering Commission, SERC since 1988, Member of SERC's Science Board since 1990.

IVERSEN, Dr Leslie, FRS, Director, Merck, Sharp & Dohme Neuroscience Research Centre since 1982. Previously Director of MRC Neurochemical Pharmacology Unit, Cambridge, 1971–1982; Locke Research Fellow of Royal Society, Department of Pharmacology, Cambridge University, 1967–1971; Harkness Fellow at Harvard Medical School, 1964–1966.

JENKINSON, Dr David, soil scientist at the Institute of Arable Crops Research, Rothamsted Experimental Station, Harpenden.

JENNINGS, Richard, Deputy Director, Wolfson Cambridge Industrial Unit, Cambridge.

JOYNER, Richard, Professor and Director of the Leverhulme Centre for Innovative Catalysis, Liverpool University.

KORNBERG, Sir Hans, FRS, Sir William Dunn Professor of Biochemistry, University of Cambridge and Fellow of Christ's College, Cambridge since 1975. Master of Christ's College, Cambridge since 1982. Member of the Cabinet's Advisory Council for Research and Development 1982–85, Member of the Agricultural and Food Research Council 1980–84, Advisory Committee on Genetic Manipulation since 1986.

KROTO, Harold, FRS, Royal Society Research Professor, Sussex University, since 1991 (Professor of Chemistry, Sussex University 1985–1991). Head, Krotographics, commercial design and graphics studio. Member of SERC Committees on Physical Chemistry, Synchrotron Radiation. Formerly Lecturer, Department of Chemistry, Sussex University 1968–1977, Reader, 1977–1985.

LADLE, Dr Mike, river ecologist at the Institute of Freshwater Ecology's River Laboratory, East Stoke, Wareham.

LEWIS, Professor Trevor, Director of Research at the Institute of Arable Crops Research, Rothamsted Experimental Station, Harpenden.

MATHIAS, Dr Peter, CBE, FBA, Master of Downing College, Cambridge, since 1987. Chichele Professor of Economic History, Oxford University, and Fellow of All Souls College, Oxford, 1969–1987. Member of the Advisory Board for the Research Councils 1983–1989.

MICHAEL, Sir Peter, CBE, Chairman of UEI plc, 1986–1989. Member of the Cabinet's Advisory Council on Research and Development (ACARD), 1982–1985. (Interviewed by the author for an earlier book, *Secrets of Success: What You Can Learn from Britain's Entrepreneurs.*)

MILSTEIN, Dr Cesar, FRS. On Scientific Staff of MRC Laboratory of Molecular Biology, Cambridge, since 1963. Currently Head, Division of Protein and Nucleic Acid Chemistry. Nobel Prize for Physiology or Medicine, jointly awarded, 1984.

MITCHELL, Dr Peter, FRS. Founder and Director of the Glynn Research Institute, Bodmin, Cornwall, 1964–1986, (Chairman 1987–1992). Director of the Chemical/Biology Unit, Department of Zoology, Edinburgh University, 1955–1963; Department of Biochemistry, Cambridge University, 1943–1955. Nobel Prize for Chemistry, 1978. Died April 1992.

MORRIS, Dr Mike, biologist, Head of Furzebrook Research Station (part of the Institute of Terrestrial Ecology).

MULVEY, Dr John, Executive Secretary of Save British Science. High energy particle physicist. Former Lecturer in Physics, Oxford University.

MUNRO, Dr Alan, Research Director and co-founder of Immunology Ltd. Previously worked in Cambridge for over 25 years in the Department of Biochemistry, the MRC Laboratory of Molecular Biology and latterly in the Department of Pathology where he was Reader in Immunology and Deputy Head of Department. Quotations based on lecture at a symposium on Innovation, December 1991, organised by the Wolfson Cambridge Industrial Unit, Cambridge.

NOBLE, Denis, FRS, Burdon Sanderson Professor of Cardiovascular Physiology, Oxford University since 1984. Member of Steering Committee, Royal Society and Fellowship of Engineering Policy Studies Unit; Foreign Secretary of The Physiological Society. Founder member of Save British Science.

O'SHEA, Michael, Director of the Neuroscience Inter-disciplinary Research Centre, Sussex University since 1991. Appointed to a London University Professorship in Cell Biology in 1988. Earlier posts include Professor of Neurobiology at Geneva University.

PADGETT, Dr Miles, Senior Consultant, Physics Group, PA Consulting Group's Cambridge Laboratory. Previously post-graduate researcher at the Cavendish Laboratory, Cambridge.

PENROSE, Roger, FRS, Rouse Ball Professor of Mathematics, Oxford University since 1973. Previously Professor of Applied Mathematics, Birkbeck College, London University 1966–1973 (Reader 1964–1966).

PERUTZ, Dr Max, OM, CH, CBE, FRS. Member of Scientific Staff of MRC Laboratory of Molecular Biology since 1979, Chairman, 1962–1979. Director, MRC Unit for Molecular Biology 1947–1962. Nobel Prize for Chemistry (jointly) 1962.

PHILLIPS, Sir David, FRS, Chairman, Advisory Board for the Research Councils since 1983; Professor of Molecular Biophysics and Fellow of Corpus Christi College, Oxford, 1966–1990, now Hon. Fellow; Research Worker, Davy Faraday Research Laboratory, Royal Institution, London, 1955–1966; Royal Society Assessor, MRC, 1978–1983; Member of ACARD, 1983–1987; Member of ACOST since 1988.

PIKE, Roy, FRS, Clerk Maxwell Professor of Theoretical Physics, King's College, London University since 1986. Previously Visiting Professor of Mathematics, Imperial College of Science and Technology, London University 1985–1986. Earlier in career worked in the Royal Signals and Radar Establishment Physics Group.

PINDER, Dr Clive, Head of the Eastern Rivers Laboratory at the Monks Wood Experimental Station.

POWLSON, Dr David, soil scientist at the Institute of Arable Crops Research, Rothamsted Experimental Station, Harpenden.

RATCLIFFE, Dr Derek, formerly Chief Scientific Officer and Deputy Director (Scientific) of the Nature Conservancy. Author of Nature Conservation in Great Britain, 1984. Retired. Currently conservation advisor to various bodies such as The National Trust.

RICH, Dr Peter, Chairman and Research Dierctor, Glynn Research Institute, Bodmin.

RICHARDS, Dr Brian, CBE. Chairman of British Biotechnology Group PLC. Chairman of the Biotechnology Joint Advisory Board. Founder member of Searle Research and Development in the UK, 1966.

RICHMOND, Sir Mark, FRS, Chairman of SERC since 1990; Member of SERC 1981–1985; Member of the Committee of Vice-Chancellors and Vice-Principals, 1987–1989; Vice-Chancellor and Professor of Molecular Microbiology, Manchester University, 1981–1990; Professor of Bacteriology, Bristol University, 1968–1981.

ROBERTS, Derek, CBE, Provost of University College, London since 1988. Joined Plessey 1953, Technical Director 1983–1985. GEC, Joint Deputy Managing Director 1985–1988.

ROBERTS, Stanley, FRSC, Professor of Organic Chemistry at Exeter University since 1986. Formerly Head of Chemical Research at Glaxo Research Group, 1980–1986, Lecturer and (from 1980) Reader in Organic Chemistry, Salford University. Member of Biotechnology Directorate Committee and SERC-funded Oxford University Interdisciplinary Research Centre.

SHARP, Margaret, Senior Fellow at the Science Policy Research Unit, University of Sussex.

SHARPEY-SCHAFER, John, Professor of Physics, Liverpool University since 1988, (Lecturer 1964–1973, Senior Lecturer 1973–1978, Reader 1978–1988). Member of SERC's Nuclear Physics Board 1986–1989.

SHEWRY, Dr Peter, Head of Long Ashton Research Station, Bristol (part of the Institute of Arable Crops Research).

SIMON, Dr Anne, technology advisor at Cazenove & Co, stockbrokers, former astronomy postgraduate at Cambridge University.

SMITH, Sir David, FRS, FRSE. Principal and Vice-Chancellor of Edinburgh University since 1987. Member of AFRC 1982–1988, SERC Science Board, 1983–1985, Co-ordinating Committee of Marine Science and Technology since 1987, Advisory Board for the Research Councils, 1989–1990.

SMITH, Desmond, FRS, FRSE, Professor of Physics and Head of Department of Physics, Heriot-Watt University, Edinburgh since 1970; Lecturer, then Reader, University of Reading, 1960–1970; Chairman of Edinburgh Instruments since 1971; Member of ACARD, 1985–1987, ACOST, 1987–1988; Member of SERC Committees 1985–1988 (Astronomy, Space and Radio, Engineering).

THOMAS, Dr Jean, FRS, Professor of Macromolecular Biochemistry, University of Cambridge since 1992, Fellow and College Lecturer, New Hall, Cambridge since 1969.

VANE, Sir John, FRS, Director of the William Harvey Research Institute since 1986, Professor of Pharmacology and Medicine, New York Medical College since 1986. Nobel Prize for Physiology or Medicine (jointly) 1982.

VENNART, Dr William, Lecturer in Physics and researcher in medical physics, Exeter University.

WEATHERALL, Sir David, MD, FRCP, FRCPE, FRS, Hon. Director, Molecular Haematology Unit, Medical Research Council, since 1980, Institute of Molecular Medicine, University of Oxford since 1988, Regius Professor of Medicine, Oxford University since 1992. President of the British Association for the Advancement of Science, 1992–1993.

WILKINSON, Sir Denys, FRS, Emeritus Professor of Physics, Sussex University, since 1987, Vice-Chancellor, 1976–1987; Professor of Experimental Physics, Oxford University, 1959–1976; Member of the British Council since 1987 (Chairman, Science Advisory Panel and Committee 1977–1986). Committee Member of CERN, 1971–1975.

WOLFENDALE, Arnold, FRS, FRAS, Professor of Physics, Durham University since 1965; Astronomer Royal since 1991. Member of SERC since 1988.

WOOD, Sir Martin, FRS, Deputy Chairman of Oxford Instruments Group plc since 1983 (founder 1959); Fellow of Wolfson College, Oxford University since 1967; Senior Research Officer, Clarendon Laboratory, Oxford University, 1956–1969; Member of ACARD 1983–1989; Member of ACOST since 1990. Director of Celltech Ltd and other companies. Reference to him is based on an interview for the author's earlier book, *Secrets of Success: What You Can Learn from Britain's Entrepreneurs*.

INDEX: PERSONS

INDEX: SUBJECTS